Global Logistics: The Connection Effect

Global Logistics: The Connection Effect

Jamie Olsen

Outline

1. .... Introduction

1

1.1.   Context, motivation and research questions

1.2.   Study outline

2. .... Research background   6

2.1.   Research in business and management

2.2.   Research philosophy

2.3.   Theoretical perspectives

2.4.   Research methods

2.5.   Research setting

3. .... Logistics outsourcing   14

3.1.   Logistics and LSP concept

3.2.   Outsourcing drivers

3.3.   Outsourcing risks

3.4.   Selection criteria and challenges

3.5.   Tender phases

3.6.   Buyer-Supplier relationships

3.7.   Boundary spanners

4. .... Guanxi   29

4.1.   Definition

4.2.   Dimensions

4.3.   Typologies

4.4.   Benefits

4.5.   Costs

4.6.   Guanxi-corruption relation

4.7.   Cultural uniqueness

5. .... Expert interviews to identify the impact of Guanxi on the LSP selection in China   45

5.1.   Guanxi of boundary spanners in the selection of LSPs

5.2.   Method: Qualitative Expert Interviews

5.2.1.   Background and evaluation

5.2.2.   Procedure

5.2.3.   Participants

5.2.4.   Analysis

5.3.   Results and Discussion

5.3.2.   Evaluation Guanxi support possibilities

5.3.3.   Evaluation Contingency factors

5.3.4.   Limitations and further research

6......Survey to verify the impact of Guanxi on the LSP selection in China   71

6.1.   The impact of Guanxi on the selection of LSPs in China

6.2.   Method: Survey

6.2.1.   Background and evaluation

6.2.2.   Procedure

6.2.3.   Participants

6.2.4.   Analysis

6.3.   Results

6.3.1.   Frequencies

6.3.2.   Crosstabs

6.3.3.   Mann-Whitney-U-Test

6.3.4.   Explorative Factor Analysis

6.3.5.   Regression Analysis

6.4.   Discussion

6.4.2.   Evaluation

6.4.3.   Limitations, implications and future research

7......Delphi study to forecast the development of the impact of Guanxi on the LSP selection in China   126

7.1.   Guanxi influence in the future

7.2.   Method: Delphi study

7.2.1.   Background and evaluation

7.2.2.   Participants

7.2.3.   Procedure

7.3.   Results

7.3.1.   Factors impacting the future Guanxi influence

7.3.2.   Future development of Guanxi influence

7.3.3.   Group differences in future Guanxi development

7.4.   Discussion

7.4.1.   Evaluation

7.4.3.   Limitations, implications and future research

8......Conclusion   164

1. Introduction

This chapter introduces the study setting and describes its outline. In section 1.1, context as well as motivation will be presented. Then, based on the identified research gap, the research questions are introduced. The study consists of in total eight chapters which will subsequently be outlined in section 1.2.

1.1. Context, motivation and research questions

Better, faster, and cheaper is the credo of the logistics sector, an industry that enables the mobility of goods and is the backbone of international trade[1]. Logistics is the "planning, implementation, and controlling of efficient, effective forward and reverse flows and storage of goods, services and related information between the point of origin and the point of consumption in order to meet customers' requirements[2]." The importance of global logistics is reflected in its market size. In 2019, it reached a value of 4.96 trillion USD and was predicted to grow further at a rate of 4.7% per year[3]. Companies in industrialized countries started in the 1980s and 1990s to outsource logistics services to address increasingly higher logistics requirements[4]. By employing dedicated Logistics Service Providers (LSPs), they can focus on their core competencies, reduce logistics costs, and have specialized logistics expertise available[5]. Usage and importance of such LSPs have been growing dramatically, and logistics is now one of the most commonly outsourced business support functions[6]. Traditionally, the relationship between a buying company and its suppliers was considered a zero-sum game, where either party's benefits came at the expense of the other[7]. These short-term, transactional buyer-supplier relationships (BSRs) then evolved to increasingly cooperative long-term partnerships, which led to financial, quality, and innovation benefits[8]. However, many cooperations fall short of expectations as little attention is given to interpersonal connections between boundary spanners, persons who operate at the organization's boundary, and manage organizational interfaces[9].

Such personal relationships are especially important in the Chinese culture. Chinese business practices are revolving around special relationships called *Guanxi*[10]. Fox Butterfield, a former foreign correspondent of the New York Times in China observed that this is a fundamentally different "mental universe" from Westerners. "We have one code of manners for all. (...) Chinese, on the other hand, instinctively divide people into those with whom they already have a fixed relationship (...) and those they don't. These connections operate like a series of invisible threads, tying Chinese to each other with far greater tensile strength than mere friendship in the West would do[11]." Contrary to the dominant Western view, where globalization and modernization would eventually lead to China's adoption of Western business practices, the

---

[1] Daugherty et al. (2011), pp. 17-18; Martí et al. (2014), p. 2982.

[2] Council of Supply Chain Management Professionals (2019).

[3] IMARC Group (2020).

[4] Daugherty et al. (2011), pp. 17-18.

[5] Kremic et al. (2006), p. 469; Sink/Langley Jr (1997), p. 182.; Wilding/Juriado (2004), p. 630.

[6] Knemeyer/Murphy (2005), p. 17.

[7] Neumann/Morgenstern (2007), pp. 46-47; Jackson (1985), pp. 120-128.

[8] Kannan/Choon Tan (2006), pp. 757-763; Terpend et al. (2008), p. 40; Carr/Pearson (1999), pp. 514-517; Larson/Kulchitsky (2000), pp. 30-36; Martin/Grbac (2003), pp. 30-36; Corsten/Felde (2005), pp. 454-457; Johnston et al. (2004), pp. 31-37.

[9] Organ (1971), p. 73; Leifer/Delbecq (1978), pp. 40-41; Katz/Kahn (1978), 17-743; Utterback (1971), p. 84; Hutt et al. (2000), pp. 51-62.

[10] Alston (1989), p. 28; Ai (2006), p. 105.

[11] Butterfield (1983), p. 44.

Chinese business culture and its different ideals and attitudes have remained over time[12]. Despite reforms in the past, in China's unique economic system, the government and the Chinese Communist Party (CCP) still play a decisive role, state-ownership is dominating, and the economy is extensively planned[13]. This leads to distortions that are also visible in China's 12.1 trillion CNY logistics market, which is coined by high costs, low efficiency, entrenched regulation, and local protectionism[14]. Since 2015, these challenges have been targeted by policy interventions aimed at upgrading the logistics network and improving its capacity and efficiency[15]. It is expected that accompanied by these developments and maturing of the market, logistics outsourcing will, just like in other countries, become more common[16]. The different institutional environment, China's wide geographical spread, large population, distinctive culture and development status have increasingly drawn the attention of researchers from the field of logistics[17]. Especially the Guanxi phenomenon has been featured as it reflects both the local institutional environment as well as local business practice[18]. For instance, studies have revealed that Guanxi impacts risk management, supplier development, and green supply chain management[19]. Also, Guanxi in the LSP context has been investigated from various angles[20]. Chu et al. proposed that Guanxi expedites service innovation of LSPs[21]. Rahman et al. found that Guanxi is one of the most critical challenges for multinational LSPs in China[22]. Moreover, Chu et al. demonstrated that Guanxi can be a dependency-coping strategy for clients of LSPs[23]. However, although it is a key topic in logistics outsourcing research, considered crucial for companies, and the importance of national culture on it has been recognized, so far, the literature overlooked the impact of Guanxi on the LSP selection process[24]. To shed light on the subject, this study attempts to answer the following research question:

RQ: How does Guanxi impact the selection of Logistics Service Providers in China now and in the future?

Derived from this overriding research question, the following research questions will be addressed sequentially and together provide an answer to the RQ. As the first study was set out to initially explore possible impacts of Guanxi on the selection of LSPs in China, its research question RQi1 states:

RQi1: How does Guanxi impact the LSP selection in China?

This study uncovered various support possibilities for members of a buying center along an LSP selection process as well as contingency factors that may alter them. To quantitatively verify or

---

[12] Tang (2015), pp. 1-317; Friedman (2012), pp. 313-332; Bell et al. (1993), pp. 1-84.

[13] Faure/Fang (2008), p. 203; European Commission (2017), p. 21; Wils-Owens (2017), pp. 4-7.

[14] Jiang, Xiao-Mei (2019), p. 8; Zhang/Figliozzi (2010), pp. 179-190; Hong et al. (2004), p. 20.

[15] Economist Intelligence Unit (2014); Jiang (2018), pp. 1-24.

[16] Chu/Wang (2012), pp. 78-79.

[17] Gong et al. (2019), pp. 548-560; Liu, Xiaohong (2014), pp. 403-405.

[18] Liu/McKinnon (2016), pp. 986-990.

[19] Abramson/Ai (1999), pp. 21-31; Cheng et al. (2012), p. 11; Lee/Humphreys (2007), pp. 454-462; Davies et al. (1995), pp. 208-213; Zhu/Sarkis (2004), pp. 282-285; Simpson et al. (2007), pp. 42-44; Luo et al. (2014), pp. 105-107; Geng et al. (2017), p. 3.

[20] Liu, Xiaohong (2014), p. 396; Huo Baofeng/Liu Chen (2017), pp. 927-928.

[21] Chu et al. (2018), p. 291.

[22] Rahman et al. (2019), pp. 607-616.

[23] Chu et al. (2019), pp. 620-621.

[24] Pagell et al. (2005), pp. 371-394; Chen et al. (2010), pp. 279-280; Chu et al. (2019), pp. 620-630; Alkhatib et al. (2015), p. 105; Aguezzoul (2014), pp. 69-78.

falsify these discoveries, research questions RQs1 and RQs2 that were addressed in the second, quantitative study state:

RQs1: Which contingency factors influence the impact of personal relationships on the LSP selection?

RQs2: How do members of a Buying center support a certain LSP during the selection process?

The results from these two studies provided insights into the current state of the impact of Guanxi on the LSP selection process. To also predict how this impact will develop in the future, a Delphi study was conducted. The guiding research questions for this forecasting method were RQd1, RQd2, RQd3a, and RQd3b that state:

RQd1: How will the influence of Guanxi on the LSP selection in China develop in the future?

RQd2: Which factors will impact how the influence of Guanxi on the LSP selection in China will develop in the future?

RQd3a: How do factors that will impact how the influence of Guanxi on the LSP selection in China will develop in the future differ among LSPs versus Buying companies?

RQd3b: How do factors that will impact how the influence of Guanxi on the LSP selection in China will develop in the future differ among Multinational versus Domestic companies?

The course of this multi-tiered investigation will be outlined in the next section[25].

1.2. Study outline

This section will describe the structure of this work. It is also illustrated in Figure 1.

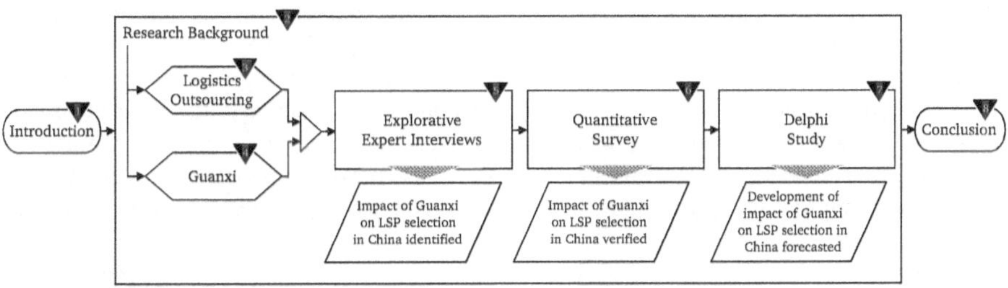

Figure 1: Study outline
Source: Own figure

Chapter 2 provides the research background of this study. First, distinctions and challenges of research in Business and Management, such as the tradeoff between theoretical rigor in relation to practical relevance, are highlighted. This is followed by an introduction to research philosophical backgrounds and the philosophical commitments made through the choice of the research approach. Then, the most prevailing perspectives and theories in research related to

Guanxi, which are Social Exchange Theory, Social Capital Theory, and the theory of Confucian Relationalism, are discussed. Subsequently, common research methods are introduced and classified based on their qualitative, quantitative or mixed method nature. Finally, the research setting, including China's automotive and logistics market, is introduced.

Chapter 3 addresses logistics outsourcing. It starts with an overview of logistics and the concept of Logistics Service Providers, including a definition of third-party logistics. Drivers, distinguished in financial and strategic/operational, and risks of outsourcing are subsequently discussed. This is followed by a synopsis of typical selection criteria employed when outsourcing logistics services. Afterwards, the six phases of a tender process for LSPs are introduced. Buyer-supplier relationships are then described, including how they can vary across a range of dimensions, what benefits and costs they can bring, and what types of such relationships exist. The chapter closes with a discussion about boundary spanners, the people who shape the cooperation between companies as linking pins.

This leads to the phenomenon of Guanxi, which is the theme of *chapter 4*. Considering that Guanxi is a multifaceted cultural phenomenon that defies an exact depiction, various definition approaches are introduced. Then, to facilitate the understanding of Guanxi's nature, the term's underlying dimensions are carved out. This is followed by a description of typologies that were developed to classify the nature of Guanxi ties. Afterwards, the benefits and costs of aggregated Guanxi, the sum of a firm's staffs' social networks and ties with other companies and government officials, are described. This is followed by a discussion about the controversial relationship between Guanxi and corruption. Finally, various perspectives about the extent to which Guanxi is universally existing in all cultures or something uniquely existing in China are introduced.

As the first section of *chapter 5* will outline, few studies assumed a micro-view on the Guanxi impact, and despite of logistics outsourcing becoming an increasingly important field in China, the effect of Guanxi on the selection of LSPs was not addressed so far. The exploratory study in the chapter aims to close this research gap by answering RQi1. To this, the technique qualitative expert interviews, a widely used method that allows gathering data with high efficiency, is introduced and evaluated. Then, the specific study procedure is explained, considering that conducting interviews about a sensitive topic in an intercultural setting requires caution concerning various points such as interview skills or language challenges. A profile of the 22 interview participants, working for LSPs and automotive Tier 1 suppliers, is afterwards provided and followed by a description of the analysis procedure. The findings in the form of support possibilities along the various stages of the LSP selection process together with contingency factors that may alter them are then summarized and discussed. The chapter concludes with an acknowledgment of the study limitations and an outlook about possible endeavors for further research.

The thought-provoking findings about the influence of Guanxi on the selection process of LSPs in China are then used as input for the study described in *chapter 6*. To verify the impact of Guanxi on the LSP selection, a survey instrument was developed and administered among supply chain management experts. For this, first, the survey method as a versatile approach for collecting information through a structured process is explained and evaluated. Subsequently, the study procedure is described, including how the questionnaire was designed and how the survey was conducted. Afterwards, the sampling approach for participants, as well as the characteristics of the survey participants, are introduced. Before the survey results are shown, the analysis approach, including a descriptive analysis and the explorative factor analysis, are described. This is followed by an outline of the questionnaire's result, which answer RQs1 and

RQs2. The chapter ends with a discussion where the findings are summarized, evaluated, and implications for theory and practice suggested.

The studies described in chapters 5 and 6 illuminate the current state of the Guanxi impact on the LSP selection process but do not assess future developments. Therefore, described in *chapter 7*, a Delphi study was conducted to address this gap and answer RQd1, RQd2, RQd3a, and RQd3b. First, the background of this method for structuring a group communication process effectively is provided and the method evaluated. Afterwards, the process for selecting qualified experts is introduced, and an overview of the participants across four panels given. This is followed by a description of the research procedure which is in line with the ranking type Delphi approach according to Schmidt. In the subsequent section about the study design, the corresponding three phases, brainstorming, narrowing down, and ranking, are explained. The results, that is the factors impacting the future Guanxi influence, the future development of the Guanxi influence, and group differences in the future Guanxi development are afterwards stated. The identified factors are then compared to the relevant research and common themes identified. At the end of the chapter, limitations and implications are highlighted, and directions for future research introduced.

Finally, in *chapter 8*, the outcomes of this whole study are summarized and implications in relation to the study context and motivation given.

2. Research background

This chapter will introduce the research background of this work. As illustrated in Figure 2, it sets the underlying framework for the study.

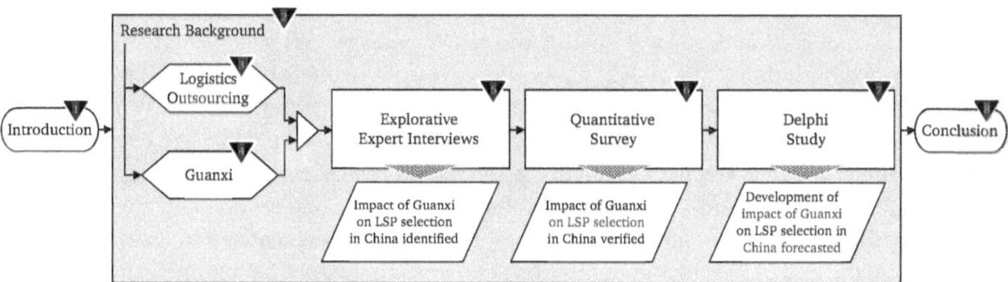

Figure 2: Chapter "Research Background" within overall research process
Source: Own figure

In section 2.1, the distinctions and challenges of business and management research, such as its multifaceted phenomena and its nature, purpose, and relevance, will be laid out. Then, in section 2.2, an introduction to research philosophy and research paradigms, world-views, or lenses through which the world is seen, is made. With these fundamentals provided, the prevailing perspectives and theories in Guanxi research, Social Exchange Theory, Social Capital Theory, and the theory of Confucian Relationalism, will be discussed throughout section 2.3. Fundamental to answer the research questions of this study is selecting appropriate research methods, which are briefly presented in section 2.4. Then in section 2.5, the research setting, the automotive supply chain and logistics market in China, is introduced.

2.1. Research in business and management

Management research tries to "develop knowledge about the management of organizations" whereby management can be defined very broadly, including aspects of companies' operations that go beyond the conventional understanding of management as planning, organizing, coordinating, and controlling[26].

Especially three factors make management and the related business research distinct. First, its phenomena are multifaceted by nature and therefore discussed by scholars across a broad spectrum of disciplines. Therefore, business and management research is highly interdisciplinary, appropriating concepts from philosophy, sociology, psychology, anthropology, economics, engineering, biology, and other disciplines with their variety of theoretical and methodological perspectives[27]. Second, managers and employees working in business tend to be relatively well educated, typically holding undergraduate, master's, or even doctorate degrees and are therefore not seldom as well educated as those researching them[28]. Third, there is an expectation among the stakeholders of business and management research, such as business students and managers, that the research has some business implications and leads to

---

[26] Alvesson/Deetz (2000), pp. 5-6. For alternative definitions and an assessment about what management research comprises, cf. Chia/Holt (2008), pp. 471-484.

[27] Antonacopoulou (2011), pp. 92-95.

[28] Easterby-Smith et al. (2015), pp. 45-127; Saunders et al. (2016), p. 6.

recommendations for management practice[29]. Hence, the nature, purpose, relevance, and utility of business and management research are disputed[30]. A particularly controversial discussion is about theoretical or methodological rigor in relation to practical relevance related to managerial knowledge. Some scholars argue that there is a fundamental gap between management researchers and practitioners as they operate based on a different logic, are engaged in different activities, and that science is too narrow, overly specialized as well as a rather closed system[31]. Other scholars believe that differences in style and language instead cause the research-practice gap, and that management researcher can generate knowledge that is both socially useful and academically rigorous[32]. Saunders et al. summarize the differences in the fundamental orientation of management researchers and practitioners. The focus of interest for researchers is basic understanding, general enlightenment, theoretical explanation, "why" knowledge, and substantive theory building, while practitioners' focus lies on usable knowledge, instrumental enlightenment, practical problem solutions, "how-to" knowledge, and local theory-in-use. The methodological imperative is theoretical or methodological rigor for researchers and timeliness for practitioners, while the key outcomes for researchers are academic publications and actionable results with practice impact for practitioners[33].

Contemporary logistics and supply chain research is criticized for favoring theory over practical relevance, which led to a "retreat into ivory towers where business academics impress each other with their erudition and give too little thought to the managerial and public policy relevance of their work[34]." This is particularly contradicting the nature of logistics problems which are typically ill-structured and from the real-world that require holistic and systematic thinking with multi-disciplinary and cross-functional approaches[35]. A study conducted in China that examined the extent to which theory-driven supply chain management (SCM) research is of practical relevance also concluded that much of the SCM research works do not translate into actionable knowledge for practitioners[36]. To respond to this critic, this study attempts to produce research that is accessible and utilizable by practitioners due to form and manner, while meeting demands for high quality and rigor[37].

2.2. Research philosophy

Research philosophy is a "system of beliefs and assumptions about the development of knowledge[38]." Researchers in business and management have to be aware of the philosophical commitments made through the choice of research strategy as it will have a significant impact on what is done and how the subject of investigation will be understood[39].

Throughout the stages of a research process, certain assumptions are made about realities that are encountered throughout the research, "ontological assumptions," human knowledge, "epistemological assumptions," and the extent and ways own values will influence the research process, "axiological assumptions[40]." Business and management research philosophies are

---

[29] Starkey/Madan (2001), S3-S5.
[30] Cassell/Lee (2011), pp. 1-12.
[31] Kieser/Leiner (2009), pp. 516-529.
[32] Hodgkinson/Rousseau (2009), pp. 534-544.
[33] Saunders et al. (2016), pp. 6-9; Saunders (2011), pp. 244-247.
[34] McKinnon (2013), pp. 10-16; Schmenner et al. (2009), pp. 339-343.
[35] Näslund (2002), pp. 323-324.
[36] Liu/McKinnon (2019), p. 76.
[37] Tranfield/Denyer (2004), p. 13.
[38] Saunders et al. (2016), p. 124.
[39] Johnson/Clark (2006), pp. iv-xxv.
[40] Crotty (1998), pp. 1-18.

scattered along a multidimensional set of continua, with the two sets of extremes being objectivism and subjectivism, that differ along the three types of assumption[41].

Ontological assumptions pose questions about what the nature of reality is and what the world is like (e.g., what are organizations like and what it is like being in an organization). In objectivism, the assumption is that reality is real, the world is externally given, and there is one true reality, universalism, which is granular and in order. In subjectivism, reality is nominal, i.e., decided by convention, the world is socially constructed, and there are multiple realities that are each chaotic and flowing[42]. *Epistemological* assumptions relate to questions such as how we can know what we know, what is considered acceptable knowledge, what constitutes good-quality data, and what kinds of contribution to knowledge can be made. Here, objectivists believe that knowledge is generated through facts and numbers, based on observable phenomena and that we can know what we know by adopting assumptions of natural scientists. Generated knowledge must be "law-like" generalizations. Subjectivists, on the contrary, trust that knowledge is generated through opinions and narratives based on attributed meanings, where assumptions of the arts and humanities are to be adopted. Knowledge is dependent on individuals, contexts, and specifics[43]. *Axiological* assumptions must address questions such as what the role of values in research is, how we should treat our own values when we do research, and how we should deal with the values of research participants. In the objectivist philosophy, research is value-free, and own values should be detached. In the subjectivist philosophy, research is value-bound and own values are integral and should be actively reflected[44].

Depending on the assumptions made, various paradigms, or worldviews through which the world is seen, can be distinguished[45]. In business and management, the paradigms positivism, critical realism, interpretivism, postmodernism, and pragmatism are typically distinguished[46]. Following the belief that researchers must present knowledge that has consequences on future applications and that management theory can never be neutral, this study takes a pragmatist view[47]. Here, the nature of reality is considered complex, rich, and external. Reality is the practical consequence of ideas, and there is a flux of processes, experiences, and practices. Acceptable knowledge is what constitutes practical meaning of knowledge in specific contexts; it focuses on problems, practices, and relevance. Research is then value-driven and initiated by a researcher's doubts and beliefs[48].

2.3. Theoretical perspectives

China is intriguing for academic researchers as the world's largest transitional economy, primary global growth driver, and country with a distinctive culture and history[49]. Especially Guanxi as a local cultural phenomenon is of broad interest for researchers in the field of economics, sociology, psychology, and business management. Also, in the China-based supply chain management literature, Guanxi is prominently featured[50]. Not least because of the attention from such a variety of disciplines, no overarching theoretical framework could be found that encompasses the wide variety of perspectives and theoretical underpinnings that

---

[41] Morgan/Smircich (1980), pp. 492-493.
[42] Saunders et al. (2016), pp. 128-131; Easterby-Smith et al. (2015), pp. 51-57.
[43] Saunders et al. (2016), pp. 128-131; Burrell/Morgan (2019), pp. 12-143.
[44] Saunders et al. (2016), pp. 128-131; Cunliffe (2003), 984-1000.
[45] Näslund (2002), pp. 323-324.
[46] Saunders et al. (2016), pp. 135-144.
[47] Kelemen/Rumens (2012), pp. 7-8.
[48] Saunders et al. (2016), pp. 135-144.
[49] Liu/McKinnon (2016), p. 973.
[50] Liu/McKinnon (2016), pp. 987-988; Chen et al. (2013), pp. 175-189; Tsui (2004), pp. 498-500.

have been employed in Guanxi research[51]. Three perspectives and theories have prevailed until today[52]. These are Social Exchange Theory, Social Capital Theory, and the theory of Confucian Relationalism[53].

According to *Social Exchange Theory* (SET), behavior can be predicted and explained by understanding the factors that individuals consider when making decisions. Like economic transactions, social exchange, "exchange of activity, tangible or intangible, and more or less rewarding or costly, between at least two parties," is made based on projections of the rewards and costs of a particular course of action[54]. However, different from strictly economic exchanges - although social exchanges also involve the principle that one person does another a favor and expects some future return – an exact nature of a return is not necessarily stipulated in advance[55]. SET has some basic assumptions[56]. a) Social behavior can be explained in terms of rewards, where rewards are goods or services, tangible or intangible, that satisfy a person's needs or goals. b) Individuals attempt maximizing rewards while minimizing losses or punishments. c) Social interaction is resulting from the fact that others control valuables or necessities and can therefore reward a person. That means to induce another to reward her or him, a person has to provide rewards to the other in return. d) Hence, social interaction is perceived as an exchange of mutually rewarding activities in which the receipt of a needed good or service is dependent on the supply of a favor in return [57]. While SET is recognized for helping to explain behavior such as issues that occur if a relationship's costs are higher than its rewards, it is also criticized from various angles. Human interaction is reduced to a purely rational process, relationships are placed in a linear structure, although some relations might adjust the degree of intimacy back and forth, and it favors openness, although full openness might not always be most suitable in relationships[58]. When applying the SET view on Guanxi, Guanxi is only exchanged after weighing costs and benefits, and reciprocity is reinforced in the social exchange[59]. The purpose of the Guanxi exchange is then to gain economic and utilitarian benefits, which means maximizing one's gains. From this theoretical lens, Guanxi practice revolves mainly around reciprocal favor exchange and gift-giving[60].

Social capital theory (SCT) explains human motivation and actions through factors such as trust, networks, and norms that go beyond economic, political, and sociological models of humans[61]. The core idea of SCT is that networks have value. Just as physical capital or human capital, i.e., connections among individuals, social capital affects the productivity of individuals and groups[62]. Social capital is the number of resources that accrue to an individual by virtue of possessing a durable network of relationships of mutual acquaintance and recognition[63]. SCT has certain characteristics[64]. a) "Social capital can be an individual or public good" and thus must be "theorized at the micro and macro level of the society[65]." b) Social capital is produced

---

[51] Chen et al. (2013), pp. 175-179.
[52] For an overview about further theories discussed in the Guanxi context, cf. Lee et al. (2018), pp. 357-358.
[53] Chen et al. (2013), pp. 175-179; Yen et al. (2017), pp. 105-107.
[54] Braithwaite/Schrodt (2015), pp. 554-555; Homans (1961), p. 13.
[55] Blau (2017), pp. 93-94.
[56] Burns (1973), pp. 188-189.
[57] Burns (1973), pp. 188-189.
[58] Miller (2005), pp. 170-186.
[59] Yen et al. (2017), pp. 105-107.
[60] Yen et al. (2017), pp. 105-107.
[61] Bartkus/Davis (2009), p. 1.
[62] Putnam (2000), pp. 18-19.
[63] Bourdieu/Wacquant (1992), p. 19. For alternative definitions, cf. Bartkus/Davis (2009), pp. 2-4.
[64] Häuberer (2011), p. 51.
[65] Häuberer (2011), p. 51.

likewise in open as well as closed structures and institutionalized and non-institutionalized relationships. These relationships can be "based on trust, authority, norms or formal organization and contain information potentials that are together the basis for access to embedded resources[66]." c) Neglected adverse effects of social capital via exclusion must be considered[67]. While some scholars argue that SCT is beneficial, as it explains how people are more successful because they are in some way better connected with other people and that one's position in a network connected to others is an asset in its own right, the theory is also subject to critique[68]. Concepts around SCT, such as trustworthiness, trust, networks, and norms, are difficult to define and quantify. Also, social capital lacks the required building blocks of "capital," for instance, it is not possible to invest in social capital at a particular time period and gain the benefits in the next time period; it cannot be sold, and it is challenging to measure it[69]. In SCT, the key principle for Guanxi is one's credibility, where Guanxi ties and networks are social resources that must be cultivated and deployed[70]. The purpose of the exchange is to attain favors and benefits and to reduce environmental uncertainty and opportunistic behavior. Guanxi practices then contain the exchange of favor as well as trust and the show and display of face (mianzi, Chinese "面子"), "a person's claimed sense of positive image in a relational context[71]."

Confucian Relationalism (CR) is based on Confucianism, which is considered the foundation of Chinese cultural values[72]. This is a distinction to the Western culture, which has its roots in ancient Greece and Rome[73]. Relationalism as a mode of governance refers to situations where the exchange conduct between committed parties is governed through relational exchange norms, i.e., shared behavioral expectations between the parties, based on the mutuality of interest[74]. Three main characteristics of CR can be abstracted[75]. a) Personal relationships are the basic unit of social systems and therefore of paramount importance. Guanxi is a deeply rooted part of Chinese culture and existing for thousands of years, where the fundamental means of control are institutionalized networks of relationships[76]. These relationships are considered as defining elements of Chinese culture and hence pursued for their own sake. Relationships therefore strongly impact all business and social life[77]. b) Personal relationships require dedication and commitment. A considerable amount of effort is needed to maintain and nurture a Guanxi relationship[78]. As relationships are developed person-to-person, they required personal interaction, i.e., spending time together[79]. Establishing and maintaining Guanxi, for example through gifts, is also costly[80]. c) Personal relationships are differentiated according to the degree of particularism; that is how close or distant they are. Closest are people where the common shared identification is based on preordained bases, like family members, relatives, or

---

[66] Häuberer (2011), p. 51.
[67] Häuberer (2011), p. 51.
[68] Burt (2005), p. 4.
[69] Bartkus/Davis (2009), pp. 5-7.
[70] Yen et al. (2017), pp. 105-107.
[71] Yen et al. (2017), pp. 105-107; Lee/Dawes (2005), p. 34.
[72] Jia/Zsidisin (2014), p. 263.
[73] Jia/Zsidisin (2014), p. 263.
[74] Bello et al. (2003), pp. 1-2; Macneil (1980), pp. 1-155.
[75] Chen et al. (2013), pp. 177-179. For a theoretical construction of the model, incl. another focus and interpretation of the characteristics, cf. Huang (2012), pp. 69-98.
[76] Chen et al. (2013), p. 193; Fei (1992), p. 29.
[77] Gold et al. (2002), p. 3; Merrilees/Miller (1999), p. 267; Wang (2007), pp. 81-86.
[78] Ewing et al. (2000), p. 84.
[79] Styles/Ambler (2003), p. 638.
[80] Fock/Woo (1998), p. 38; Yi/Ellis (2000), pp. 28-29.

members of the same clan[81]. More distant are relationships with people based on a shared social identity, such as the identification with the same hometown, workplace, or based on a common third party that acquaints them [82]. In the theory of CR, Guanxi refers to "all personal relationships, signifying the humaneness of a person and a society[83]." The purpose of an exchange is to behave ethically as a personal commitment to relationship appropriateness. Guanxi practice is to engage in various social norms considered appropriate and humane[84].

Assuming a SET or SCT perspective on Guanxi has some caveats. Those theories do not acknowledge the "corrosive downside" of Guanxi that resembles corrupt activities (cf. section 4.6)[85]. They also do not reflect that Guanxi transactions also require managing a network of resource coalitions and operating within a hierarchical structure[86]. Finally, they fail to recognize the Guanxi-inherent element of emotional bonding through social interactions[87]. Therefore, while acknowledging the ongoing debate in the literature, this study follows the perspective of Confucian Relationalism[88].

2.4. Research methods

Research methods are "techniques and procedures used to obtain and analyze data[89]." They can be distinguished whether they are following a qualitative, quantitative, or mixed methods approach[90].

In the *qualitative research* approach, open research questions are examined in detail on few units using unstructured or partially structured data collection methods. The goal is to get an in-depth understanding of an object and to build theory. The researcher is actively involved as a participant or catalyst. Sample sizes are typically small, and the sample design can be purposive and nonprobabilistic[91]. Here, data can be collected faster due to the limited sample size, which allows for a shorter feedback turnaround. Also, as insights are developed while the research progresses, data analysis can be shortened. Collected qualitative, i.e., non-numerical, verbal, and visual, data are analyzed via a human analysis, following computer or human coding, and the process is always ongoing during the project. The research design may evolve or adjust during the project, and consistency is not expected[92]. Qualitative methods include expert interviews, participant observations, and ethnographies, document analysis, or case studies[93].

In the *quantitative research* approach, theoretically derived research hypotheses are examined on many units with structured data collection methods. The goal is typically to describe, predict, or to build and to test a theory. The sample design is probabilistic, and the sample size large, which lengthens the data collection. Collected quantitative, i.e., numerical, data are analyzed with statistical and mathematical methods[94]. The analysis may be ongoing during the project, and a clear distinction between facts and judgments is maintained. Insights and meaning are

---

[81] Tsang (1998), p. 65.
[82] Chen et al. (2013), pp. 171-172; Chen/Chen (2004), pp. 311-312.
[83] Yen et al. (2017), pp. 105-107.
[84] Yen et al. (2017), pp. 105-107.
[85] Li (2011), p. 3; Fock/Woo (1998), p. 38.
[86] Su et al. (2007), pp. 303-316.
[87] Chen et al. (2013), pp. 167-207; Yen et al. (2017), pp. 105-106.
[88] Yen et al. (2017), pp. 105-106.
[89] Saunders et al. (2016), pp. 3-4.
[90] Döring et al. (2015), pp. 184-185.
[91] Döring et al. (2015), pp. 184-185; Cooper/Schindler (2014), pp. 146-148.
[92] Döring et al. (2015), pp. 184-185; Cooper/Schindler (2014), pp. 146-148.
[93] Cooper/Schindler (2014), p. 130.
[94] Cooper/Schindler (2014), pp. 146-148; Döring et al. (2015), pp. 184-185.

limited by the opportunity to probe respondents and the original data collection instrument's quality. Here, the research design is determined before commencing the project, and consistency is critical[95]. Quantitative methods include survey research, experimental research, and psychometric tests[96].

Different from traditional, mono-method designs that follow either a purely qualitative or purely quantitative approach, in *mixed methods research*, qualitative and quantitative approaches are combined in the context of a single study. Typically, qualitative and quantitative sub-studies that are related to each other are conducted in parallel or subsequently. For instance, in the "pre-study" mixed method design, a qualitative study serves as a preliminary study to generate hypotheses that are then tested in a quantitative study[97]. Other reasons for using mixed methods designs might be that during the research, one method may lead to the discovery of new insights which are followed up through the use of the other method, or that meanings and findings are elaborated, enhanced, clarified, or linked[98]. Mixed method research can also be used to increase the validity of a construct, enable interpretability, meaningfulness, and validity of constructs or increase the breadth and depth of inquiry results[99]. Various combinations of qualitative and quantitative methods are conceivable[100]. A special case is the Delphi method which can be, depending on the used type of Delphi, considered a mixed method itself[101].

2.5. Research setting

This section will describe this study's research setting, which is China's automotive and logistics market.

China's automotive market

As each car contains over 20,000 parts, typically sourced from thousands of suppliers from various countries, the automotive supply chain is considered one of the world's most complex[102]. Tier 1 suppliers, companies that supply parts or systems directly to original equipment manufacturer (OEM), are a significant driver of innovation and account for 60-70% of a new car's production costs[103]. This industry produces an expensive and intricate product, which is capital intensive and time-consuming to develop[104]. However, today's customers demand increasingly new functionalities in cars, and the adoption of new technologies is driving change[105]. The consultancy Oliver Wyman suggests that this will lead to a disruption of the supply chain, which will end transaction-based purchasing and lead to restructuring and consolidation of 50% of suppliers[106].

China is the largest producer and largest consumer market for cars. In 2019, 25.7 million vehicles were built locally, more than in the next largest countries USA (10.9 million), Japan

---

[95] Cooper/Schindler (2014), pp. 146-148; Döring et al. (2015), pp. 184-185.
[96] Döring et al. (2015), p. 15.
[97] Döring et al. (2015), pp. 184-185.
[98] Saunders et al. (2016), p. 173.
[99] Greene et al. (1989), p. 259.
[100] Greene et al. (1989), pp. 255-272.
[101] Tapio et al. (2011), pp. 1618-1627; Kennedy (2004), pp. 504-506.
[102] Kapadia (2018).
[103] Scannell et al. (2000), p. 23; Thun (2018), pp. 7-18.
[104] Saranga et al. (2019), p. 101495.
[105] Kaas et al. (2016), pp. 3-16.
[106] Joas et al. (2019), pp. 7-8.

(9.7 million), and Germany (4.7 million) combined[107]. Thanks to smart industry policies and strong transformation efforts, China could create a strong, booming automotive supply chain in China since its opening in the 1980s [108]. The country's automotive industry is mostly concentrated across six main clusters in Jilin, Beijing, Hubei, Guangdong, Chongqing, and the Yangtze river delta[109]. In the last decade, China's vehicle production proliferated at a rate of some 15% per year[110]. Recently, however, the market started to show signs of cooldown so that in 2019, 7.5% fewer vehicles were produced than in the year before[111].

China's logistics market

Despite its position as the world's leading exporter, China suffers from an inefficient logistics market[112]. According to Hung Lau and Zhang, this is due to especially three key challenges[113]. *Poor infrastructure*, a lack of adequate transportation networks, IT infrastructure, and little integration of transportation, warehousing, and distribution facilities, which leads to higher transportation costs and a lack of reliability in pick-up and delivery time[114]. *Regulation issues*, i.e., local protection, have restrained the development of national service networks and made it difficult for LSPs in China to fully meet their clients' requirements. *Finding qualified staff*, a lack of logistics training programs and a lack of high-quality providers with the scope and scale to fully meet customers' requirements, which gives shippers little confidence in the service levels of LSPs and makes it difficult to find good providers that can deliver high quality and consistent services across geographical regions[115]. Other scholars recently identified low efficiency, high logistics cost, congestion, lack of a nationally integrated intermodal transport network, entrenched regulation and local protectionism, poor IT infrastructure, and inability to use advanced technology as well as underdeveloped warehousing service as crucial issues in China's logistics market[116]. However, in 2013 government initiatives aimed at improving China's logistics network and its capacity were launched, and over the following years, the logistics industry could improve its quality, efficiency, and innovativeness[117]. The share of logistics costs in relation to GDP in China was 14.6% in 2017, which is a significant improvement compared to the 16.9% in 2013, however it still almost double the value of major developed countries, such as Germany or the United States (US) at 8%[118].

---

[107] International Organization of Motor Vehicle Manufacturers (2020).

[108] Saranga et al. (2019), p. 101498.

[109] Scherer/Ng (2014), p. 11.

[110] International Organization of Motor Vehicle Manufacturers (2009); International Organization of Motor Vehicle Manufacturers (2019).

[111] International Organization of Motor Vehicle Manufacturers (2020).

[112] Economist (2014).

[113] Hung Lau/Zhang (2006), p. 780.

[114] Hung Lau/Zhang (2006), p. 780.

[115] Hung Lau/Zhang (2006), p. 780.

[116] Zhang/Figliozzi (2010), pp. 179-190; Hong et al. (2004), p. 20.

[117] Economist Intelligence Unit (2014); Jiang (2018), pp. 1-24.

[118] Jiang, Xiao-Mei (2019), p. 8; Fang et al. (2020), p. 4; Armstrong & Associates (2020).

3. Logistics outsourcing

In this chapter, the topic of logistics outsourcing will be reviewed. As illustrated in Figure 3, this is the first key concept that lays the foundation for the subsequent explorative study.

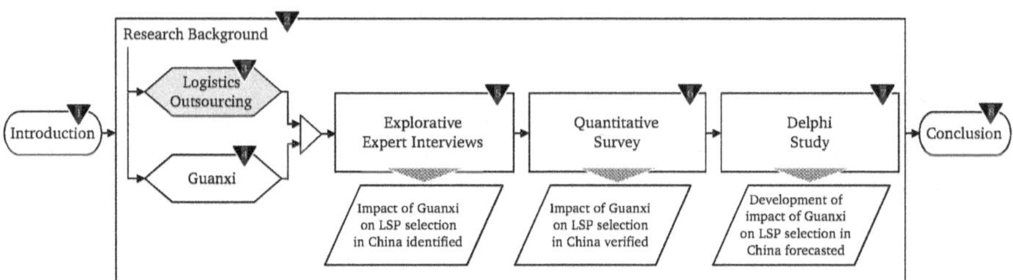

Figure 3: Chapter "Logistics Outsourcing" within overall research process
Source: Own figure

Initially, an overview of logistics and third-party logistics, the outsourcing of logistics services to an LSP, will be given in section 3.1. This is followed in section 3.2. by a catalog of financial and strategic/operational reasons that drive companies to outsource the function. While there are several benefits from outsourcing logistics services, companies also face certain risks, which are subsequently discussed in section 3.3. If companies ultimately have opted to employ an LSP, they must decide for one. Commonly used selection criteria, as well as related challenges when procuring logistics services, are therefore discussed in section 3.4. Afterwards, the six typical phases of a tender process for LSPs are introduced in section 3.5. Central to unlocking the benefits of outsourcing is the effective management of buyer-supplier relationships. Therefore, they are described in section 3.6, including how they can vary across various dimensions, what benefits and costs they can bring, and what types exist. Boundary spanners, the persons who operate at the periphery of an organization, manage these BSRs. The types of functions that boundary spanners perform and their challenges in their role are then discussed in the final section 3.7.

3.1. Logistics and LSP concept

In this section, definitions for the term *Logistics* as well as *LSP*, the provider that carries out outsourced logistics activities, will be provided.

Logistics

Logistics includes all activities that facilitate product movement and the coordination of supply and demand and add value to a firm's or channel's output by creating place, time, and quantity utilities for the customer[119]. It involves getting the *right product* in the *right way* in the *right quantity and right quality* in the *right place at the right time*, for the *right customer* at the *right cost*[120]. The most common definitions of the term are flow oriented[121]. For instance, the Council

---

[119] Cf. Pfohl (1997), p. 153; Pfohl (1972), pp. 1-217.
[120] Mangan (2011), p. 27.
[121] For further definition approaches, cf. Pfohl (2018), pp. 3-21 and Lummus et al. (2001), pp. 426-431.

of Supply Chain Management Professionals (CSCMP) defines the management of logistics as "the part of supply chain management that plans, implements, and controls the efficient, effective forward and reverses flow and storage of goods, services and related information between the point of origin and the point of consumption in order to meet customers' requirements[122]." It is delineated from supply chain management, that according to the CSCMP "encompasses the planning and management of all activities involved in sourcing and procurement, conversion, and all logistics management activities. Importantly, it also includes coordination and collaboration with channel partners, which can be suppliers, intermediaries, third party service providers, and customers. In essence, supply chain management integrates supply and demand management within and across companies[123]." Also, the European Logistics Association (ELA) employs a flow-based definition and considers logistics as the "organization, planning, control, and execution of the goods flow from development and purchasing, through production and distribution, to the final customer in order to satisfy the requirements of the market at minimum costs and minimum capital use[124]." Fueled by increasing economic development and trade, the global logistics market expanded continuously and reached a value of 4.96 trillion USD in 2019[125].

Companies today recognize that logistics capabilities can make a major contribution to the overall performance and be a source of competitive advantage by creating a differentiated customer value[126]. However, various company external and internal forces are transforming logistics and pressure companies to find appropriate ways to adapt[127]. Price transparency and growing competition increases cost pressure. Rising customer requests for individualized goods require a more diverse product portfolio and diversified logistics services[128]. Flexible structures in product creation and complementary logistics processes are needed to compensate customer's increasingly fluctuating demand. Demographic change and a transformed competence profile exacerbate logistics talent shortage. New government regulations force companies to comply swiftly to changed laws, guidelines, and duties[129]. Moreover, the digitalization of business processes requires defined interfaces or uniform systems to enable data exchange across all stages of complex value creation networks[130].

LSP

A way to cope with these increasing logistics requirements is to outsource logistics services to specialized LSPs[131]. Tasks previously done internally are shifted to an external provider, often accompanied by a transfer of employees[132]. Outsourcing in a logistics context is frequently referred to as "third-party logistics" (TPL), carried out by LSPs, although terms such as logistics alliance or contract logistics are also used[133]. While these terms are customary employed, no single, consistent definition of the concepts prevails[134]. *Narrow* definitions of TPL link it to

---

[122] Council of Supply Chain Management Professionals (2019).
[123] Council of Supply Chain Management Professionals (2019).
[124] European Logistics Association (1993).
[125] IMARC Group (2020).
[126] Morash et al. (1996), pp. 1-2.
[127] Mazur et al. (2019), p. 4.
[128] Kersten et al. (2017), pp. 14-20.
[129] Kersten et al. (2017), pp. 14-20.
[130] Kersten et al. (2017), pp. 14-20.
[131] Daugherty et al. (2011), pp. 17-18; Kremic et al. (2006), p. 469.
[132] Belcourt (2006), p. 270.
[133] Bagchi/Virum (1996), p. 93; Alkhatib et al. (2015), pp. 103-105.
[134] Skjoett-Larsen (2000), pp. 113-115.

distinctive functional and/or interorganizational features of the outsourcing relationship[135]." For example, Berglund et al. define TPL as "activities carried out by a Logistics Service Provider on behalf of a shipper and consisting of at least management and execution of transportation and warehousing (if warehousing is part of the process). Besides, other activities can be included, for example, inventory management, information-related activities, such as tracking and tracing, value added activities, such as secondary assembly and installation of products, or even supply chain management [136] ." They also require the contract to "contain some management, analytical or design activities, and the length of the cooperation between shipper and provider to be at least one year, to distinguish third-party logistics from traditional 'arm's length' sourcing of transportation and/or warehousing[137]." They further emphasize that their definition of TPL excludes the execution of basic transportation activities that are not carried out in combination with warehousing and other activities[138]. *Broad* definitions consider TPL as any form of outsourcing of logistics activities previously performed in-house[139]. One example is Lieb et al.'s definition, where TPL is the "use of external companies to perform logistics functions which have traditionally been performed within an organization. The functions performed by the third-party firm can encompass the entire logistics process or selective activities within that process[140]." Another is Coyle et al.'s, according to which TPL involves an "external organization that performs all or part of a company's logistics functions[141]." Other definitions focus on the requirement to provide multiple services so that TPL services are seen as "multiple distribution activities provided by an external party, assuming no ownership of inventory, to accomplish related functions that are not desired to be rendered and/or managed by the purchasing organization[142]." Alternative definitions refer to the relationship aspect and consider TPL as a "close and long-term relationship between a customer and a provider encompassing the delivery of a wide array of logistics needs [where] parties ideally consider each other as partners. (...) The goal of the relationship is to develop a win-win arrangement[143]." The European Union (EU) project PROTRANS integrates these features and defines TPL as "activities carried out by an external company on behalf of a shipper and consisting of at least the provision of management of multiple logistics services. These activities are offered in an integrated way, not on a standalone basis. The co-operation between the shipper and the external company is an intended continuous relationship[144]." Due to its integrative character, the latter definition of TPL will also be adopted in this work, and hence an LSP is referred to as an external company carrying out such activities.

Functions provided by LSPs today include (percentages indicate how often shippers in the US outsource this function) domestic transportation (81%), international transportation (71%),

---

[135] Marasco (2008), pp. 128-129.

[136] Berglund et al. (1999), pp. 59-60.

[137] Berglund et al. (1999), pp. 59-60.

[138] Berglund et al. (1999), pp. 59-60.

[139] Marasco (2008), pp. 128-129. The difference between narrow and broad definitions might be related to the evolution of the sector, which emerged from service providers that previously offered warehousing and transport services and later added new and unique value-added services, cf. Bowersox/Closs (1996), pp. 110-115. In the 1980s, traditional Logistics Service Providers with a historically strong position in warehousing or transport developed into LSPs. Then, in the early 1990s, network players, parcel and express companies started TPL activities, grounded in their global air express networks and experience with expedited freight. Finally, in the late 1990s, companies from non-traditional logistics fields such as IT, management consulting or financial service entered the market (e.g. IBM, Microsoft, GE Capital Services), Berglund et al. (1999), pp. 62-65. As next development step, now fully digital forwarders emerge, cf. Elbert/Gleser (2019), pp. 19-31.

[140] Lieb et al. (1993), p. 35.

[141] Coyle et al. (2003), p. 425.

[142] Sink et al. (1996), p. 40.

[143] Bagchi/Virum (1996), p. 95.

[144] Buck Consultants International (2003), p. 2.

warehousing (69%), freight forwarding (50%), customs brokerage (40%), freight bill auditing and payment (35%), cross-docking (28%), transportation planning and management (28%), product labeling, packaging, assembly, kitting (26%), reverse logistics (defective, repair, return) (24%), inventory management (22%), order management and fulfillment (19%), supply chain consulting services (14%), IT services (11%), fleet management (11%), service parts logistics (9%), 4PL services (9%) and customer service (6%)[145].

3.2. Outsourcing drivers

Outsourcing drivers, reasons why companies perform outsourcing, can be roughly distinguished across the financial and strategic/operational categories.

Financial

Most organizations want to achieve cost savings through outsourcing[146]. This is theoretically attainable if the costs for a supplier's service are lower than the own organization's, incorporating the supplier's additional overhead and profit, as well as transaction costs[147]. Such levels of efficiency can be achieved through specialization and economies of scale[148]. Already the economist Adam Smith hypothesized in 1776 the benefits of specialization, resulting in absolute gains, which was later repeatedly confirmed[149]. Therefore, companies and countries now typically focus more on producing a relatively narrow range of goods and services. In specific niches, they have developed "economies of scale, scope, and knowledge intensity so formidable that neither smaller nor more integrated producers can effectively compete with them[150]." Companies that outsource can benefit from lower costs, especially in the short term, as less fixed asset investments are required, and less indirect costs for support systems might be needed[151]. Other cost-related benefits of outsourcing include a transfer of fixed to variable costs and perceived increased cost control[152].

Also, considering logistics outsourcing, the main driver for companies is to reduce logistics costs[153]. Although this factor has always been among the top reasons, over time, its relative importance decreased[154]. Other important financial-related factors are the possibilities to increase efficiency and utilize assets better[155]. Besides, capital expenditures as well as the employee base, including the related costs for recruitment, selection, and training, can be reduced[156]. With less own logistics assets, equipment maintenance costs and the financial risk can be minimized[157]. Outsourcing of logistics activities also happens when merger/acquisition

---

[145] Langley Jr (2019), p. 9. A 4PL is "an independent, singularly accountable, non asset based integrator of a clients supply and demand chains. The 4PL's role is to implement and manage a value creating business solution through control of time and place utilities and influence on form and possession utilities within the client organisation. Performance and success of the 4PL's intervention is measured as a function of value creation within the client organization," Win (2008), p. 677.

[146] Belcourt (2006), p. 270; Loh/Venkatraman (1992), pp. 7-24; Holcomb/Hitt (2007), pp. 464-465.

[147] Harler (2000), p. 56; Bers (1992), pp. 54-57. For an introduction about transaction costs, cf. Williamson (1981), pp. 548-577.;

[148] Kremic et al. (2006), pp. 468-469.

[149] Smith (1776), pp. 1-7; Ethier (1982), pp. 403-404.

[150] Hummels et al. (1998), p. 81; Quinn/Hilmer (1994), p. 51.

[151] Muscato (1998), pp. 8-11; Hubbard (1993), p. 46; Kremic et al. (2006), p. 468; Gilley/Rasheed (2000), p. 765.

[152] Blumberg (1998), p. 7; Alexander/Young (1996), pp. 116-119.

[153] Wilding/Juriado (2004), p. 630; Wilding and Juriado showed that this is different for consumer goods companies that outsource primarily to benefit from the competencies of 3PLs, cf. Wilding/Juriado (2004), pp. 635-636.

[154] Alkhatib et al. (2015), p. 111; Langley Jr (2015), p. 12.

[155] Wilding/Juriado (2004), p. 630.

[156] Sink/Langley Jr (1997), p. 182; van Damme/Amstel (1996), p. 88.

[157] Selviaridis et al. (2008), p. 385.

activities require keeping assets off the books or management demands for a financial contribution from all company sectors[158].

Strategic/Operational

Another crucial driver of outsourcing is a company's need to focus on core competencies and increase flexibility[159]. Core competencies are activities that offer a long-term competitive advantage, i.e., have the highest degree of contribution to a company's competitiveness[160]. Competition forces companies to redirect scarce resources to these competencies to maximize resource efficiency and gain a competitive edge[161]. Another critical driver of outsourcing, caused by an increasingly volatile, uncertain, complex, and ambiguous market environment is flexibility, "the ability to change or react with little penalty in time, effort, cost or performance[162]." Outsourcing helps to acquire the ability to respond to changes in customer demands in scale and scope[163]. In addition, external suppliers' investments, innovations, and capabilities, which would otherwise be too expensive to replicate, can be gained[164]. By accessing the skills of sophisticated, specialized suppliers, higher quality results can be achieved than an individual company could[165].

By employing an LSP that can offer customized supply chain solutions, service levels (e.g., lead time or the number of order cycles), and customer satisfaction can ultimately be improved[166]. Another important driver is the possibility of having greater and more specialized logistics expertise available while being able to focus on core competencies[167]. Through outsourcing and the related tender processes, valuable information for logistics performance management, logistics improvement, and development, as well as cost benchmarking, can be gained[168]. An LSP also provides access to logistics information technology systems, e.g., for handling the information flow loop, and typically has more logistics assets and partnerships available, which gives its customers volume flexibility and access to global logistics networks[169]. This becomes especially important in uncertain markets coined by increasing globalization, market recessions, and sustainability issues[170]. Using LSPs can also reduce the supply chain risk, as risk is shared over a wider base of storage points[171]. Supply chain risk refers to "anything that presents a risk (i.e., an impediment or hazard) to information, material and product flows from original suppliers to the delivery of the final product to the ultimate end-user[172]." Amplifiers for the risk reduction effect are a complex supply chain due to a fragmented supply base, a high volume of product returns, and complex supply processes[173]. Outsourcing tends to occur when inventory needs to be moved faster, during restructuring activities, or to determine a product's competitive advantage in the marketplace[174]. Through the usage of a local LSP, a shipper can

---

[158] Muller (1992), pp. 60-67.
[159] Kremic et al. (2006), p. 469.
[160] Quinn/Hilmer (1994), p. 43; Prahalad/Hamel (1990), pp. 79-91.
[161] Drtina (1994), pp. 56-62; Ketler/Walstrom (1993), pp. 449-459.
[162] Upton (1994), p. 73; Bennett/Lemoine (2014), p. 313; Antonucci et al. (1998), pp. 26-31.
[163] Kamien/Li (1990), pp. 1352-1363; Gupta/Gupta (1992), p. 47; Pagnoncelli (1993), pp. 15-22.
[164] Kakabadse/Kakabadse (2000), p. 690.
[165] Quinn/Hilmer (1994), p. 51; Kremic et al. (2006), p. 469.
[166] Bask (2001), pp. 484-486.
[167] Sink/Langley Jr (1997), p. 182.
[168] Alkhatib et al. (2015), p. 116.
[169] Richardson (1990), pp. 17-20; Selviaridis et al. (2008), p. 385.
[170] Alkhatib et al. (2015), p. 103.
[171] Bolumole (2001), pp. 93-94.
[172] Peck (2006), p. 132. For a discussion about alternative definitions of supply chain risk, cf. Heckmann et al. (2015), p. 122.
[173] Bolumole (2001), pp. 93-94.
[174] Muller (1992), pp. 60-67; Byrne (1993), pp. 58-62.

also create a local image in a market without investing in local assets, which becomes particularly relevant when expansion into unfamiliar markets is happening[175]. Due to the country's vast size, this is especially relevant for China, where the typically young and fast-developing local companies employ LSPs to facilitate a quick market penetration to secure market shares[176].

3.3. Outsourcing risks

Following outsourcing operations, an outsourcing company might face certain negative consequences[177]. Unsurprisingly, the decision to outsource leads to a risk of dependence on the vendor. Often there is no safety net if a vendor fails to perform as expected[178]. This might occur due to higher costs, service levels that do not meet expectations, inadequate quality, a loss of competitive advantage, or opportunistic behavior[179]. The dependence becomes particularly crucial as after the outsourcing activity is conducted, the power in the relationship shifts to the supplier, and control might be difficult or is lost[180]. While changes in the economic and technological setting likely occur, outsourcing relationships might stay relatively rigid, which leads to an actual reduction in flexibility instead of the expected increase[181]. Such changes are hard to predict and accounted for in contracts, which explains why several studies found that managers complain about contracts that are too unwieldy and inflexible to ensure sustained value[182]. Through outsourcing, an organization also faces the risk of losing certain critical skills or knowledge that might be difficult to reacquire[183]. This might lead to a loss of innovation capabilities, opportunities, and ultimately customers[184].

Buyers of LSP services perceive a risk of reduced control over activities, cost increase, customer non-acceptance, loss of in-house logistics expertise, increased supplier dependence, as well as impaired credibility with customers[185]. LSPs might suffer from issues related to their workforce, such as strike, labor union disputes, but also a lack of experience or training, as well as issues related to their IT systems[186]. They might also stop their service as they exploit changes in the supplier market or become insolvent[187]. Such concerns are, for example, reflected in the results of a survey among users of LSPs, which revealed that after employing an LSP, time and effort spent on logistics have not decreased, that the quality of third-party employees remained disappointing, that the transition during the implementation stage was unsatisfactory, or that customer complaints have increased[188]. Speed and delivery reliability might be more effectively realized by maintaining the function in-house, as suppliers could be unable to handle volume demand changes or delivery requirements[189]. In China, companies also mention a lack of LSP

[175] Bradley (1994a), 56 A3-A14; Bradley (1994b), pp. 66-71; Maltz (1995), pp. 73-80.
[176] Hung Lau/Zhang (2006), p. 787.
[177] Aubert et al. (1998), pp. 685-690.
[178] Quélin/Duhamel (2003), pp. 656-657.
[179] Gordon/Walsh (1997), p. 273; Kakabadse/Kakabadse (2000), p. 713; Prahalad/Hamel (1990), pp. 79-91; Quinn/Hilmer (1994), pp. 52-55; Hätönen/Eriksson (2009), pp. 142-155.
[180] Antonucci et al. (1998), pp. 26-31.
[181] Gordon/Walsh (1997), p. 274; Kremic et al. (2006), p. 469.
[182] Bryce/Useem (1998), p. 639.
[183] Campbell (1995), p. 22.
[184] Quinn/Hilmer (1994), pp. 52-55; Roberts (2001), pp. 239-249.
[185] Sink et al. (1996), p. 43.
[186] Wu et al. (2006), pp. 353-360; Ho et al. (2015), pp. 5039-5044.
[187] Chopra/Sodhi (2004), p. 54; Cooke (2002), p. 31.
[188] Sink/Langley Jr (1997), p. 182; The last point could be related to the fact that logistics outsourcing will reduce the frequency of the customer contact which in turn could mean that customers feedback reaches a company with delay, cf. van Damme/Amstel (1996), p. 89.
[189] Walker/Weber (1984), pp. 383-390; van Damme/Amstel (1996), p. 89; Noordewier et al. (1990), pp. 80-93.

capability, inadequate transportation and IT infrastructure, and an absence of overall post-outsourcing reviews[190].

3.4. Selection criteria and challenges

Once companies decided to employ an LSP, they have to select one based on certain criteria. These will be discussed in this section, as well as challenges when selecting LSPs pointed out.

Criteria

Various studies identified that typically numerous tangible and intangible criteria are considered in the LSP selection process[191]. For instance, Spencer et al. listed 23 evaluative criteria for choosing outside service vendors with on-time performance, service quality, good communication, reliability, and service speed as the most important ones[192]. Leahy et al. identified 25 determinants of successful Third-Party relationships ranging from access of parties to the latest technology, which allows the buyer to use the provider's latest technology and equipment without the burden of financial investment, over control and performance appraisal, to cost savings and customer orientation[193]. Verman and Pullmann recognized the four factors on-time delivery, cost, delivery lead time, and flexibility[194]. In a literature review about selection criteria in articles published in 1994-2013, Aguezzoul could classify the 11 key criteria cost, relationship, services, quality, information & equipment system, flexibility, delivery, professionalism, financial position, location, and reputation[195]. Alkhatib et al. also conducted a literature review to identify the most used selection criteria in articles published between 2008-2013. They found that cost/price, quality, and reliability, flexibility and compatibility, services, financial measures, sustainability measures as well as delivery were the most important ones[196]. Further identified criteria were IT, management and organization, risk, geographical location, reputation and status, relationship and collaborations as well as global abilities[197]. They also compared their results with other literature reviews conducted in 1996-1991 by Weber et al. and 2000-2008 by Ho et al., and could demonstrate that the relative importance of criteria changed over time[198]. While for example after 2008, cost and price were the most important criteria, quality was more important during the 1990s[199].

Challenges

In contrast to goods, procuring services such as selecting LSPs is considered more difficult and more expensive[200]. This is typically attributed to the four features of services, intangibility, heterogeneity, inseparability, and perishability[201]. *Intangibility* is considered the key difference between a good and a service. However, it is also recognized very few offerings are purely intangible or totally tangible, but instead, that services tend to be more intangible than goods,

[190] Hung Lau/Zhang (2006), p. 787.
[191] Aguezzoul (2014), pp. 69-78.
[192] Spencer et al. (1994), pp. 70-71.
[193] Leahy et al. (1995), pp. 12-13.
[194] Verma/Pullman (1998), pp. 739-750.
[195] Aguezzoul (2014), pp. 69-78.
[196] Alkhatib et al. (2015), p. 133.
[197] Alkhatib et al. (2015), p. 133.
[198] Weber et al. (1991), pp. 2-18; Ho et al. (2010), pp. 16-24; Alkhatib et al. (2015), p. 111.
[199] Alkhatib et al. (2015), p. 111.
[200] Ellram et al. (2007), p. 47; Wynstra et al. (2018), pp. 90-92.
[201] Lovelock (1983), pp. 10-19.

and goods tend to be more tangible than services[202]. The lack of physical attributes renders it challenging to define precise specifications, pre-determine costs, and to verify bills respectively that a contract is completed[203]. *Heterogeneity* is the recognition that a service cannot be duplicated exactly for each customer, often due to the required tailoring to diverse organizational needs, especially for production-related services[204]. This makes the definition and measurement of quality, although it is recognized as paramount, subjective, and problematic[205]. *Inseparability*, or "co-production," reflects the essential involvement of customers in the service delivery[206]. Production and consumption of a service are often occurring simultaneously and in proximity, leading to an increased number of interactions[207]. As the customer's input in the service exchange co-determines the service outcome, an exact supplier's performance measurement is problematic[208]. *Perishability* implies that it is not possible to "inventory a service." If the demand for a service exceeds supply on a certain day, the excess business is lost, while if the supply exceeds demand, the oversupply is lost[209]. Therefore, demand fluctuations must be buffered with additional capacity, and frequent communication between providers and customers is needed[210]. The actual challenges of procuring services depend mainly on the writing of specifications, respectively statements of work[211]. If this is not properly done, it becomes challenging to determine service levels, and hence understand what must be measured subsequently and how[212]. To address this challenge, Axelsson and Wynstra propose four service definition methods. *Input specifications* focus on the resources and capabilities of the supplier, *throughput specifications* focus on the supplier processes needed to produce the service, *output specifications* focus on the functionality, or the performance of the service, and *outcome specifications* focus on the economic value for the customer to be generated by the service[213]. Van der Valk and Rozemeijer, therefore, suggest that the purchasing process for a service should include the two additional steps a) request for information and b) detailed specification development, based on the information obtained from the request for information step in interaction with the suppliers, before a supplier gets selected[214].

3.5. Tender phases

The tender process for LSPs is a multi-dimensional decision of high importance for companies[215]. It is similar to the general purchasing process and typically occurs in six phases[216]. *Phase 0: Project start and acquisition.* After the Industrial Client (IC) decides to implement a contract logistics project, a related internal project management process starts with a kick-off

---

[202] Zeithaml et al. (2018), 6-7.
[203] Ellram et al. (2007), p. 48.
[204] Jackson et al. (1995), pp. 100-106; Fitzsimmons et al. (1998), pp. 370-379.
[205] Pleger Bebko (2000), pp. 9-10.
[206] Lovelock/Young (1979), pp. 168-178.
[207] Chathoth et al. (2013), pp. 12-14.
[208] Nullmeier et al. (2016), pp. 30-34.
[209] Lovelock (1983), pp. 16-17.
[210] Ellram et al. (2007), p. 48.
[211] Smeltzer/Ogden (2002), pp. 63-69.
[212] Van der Valk/Rozemeijer (2009), pp. 6-7.
[213] Axelsson/Wynstra (2002), pp. 143-144; van Weele (2010), pp. 96-98.
[214] Van der Valk/Rozemeijer (2009), pp. 5-7.
[215] Alkhatib et al. (2015), p. 105; Aguezzoul (2014), pp. 69-78.
[216] Pfohl (2016), pp. 345-346. Although a variety of models for a general purchasing process exist, for instance Day/Barksdale (1994), pp. 45-51 or Ellram et al. (2007), p. 47, they typically come down to six steps. 1. Define specifications, 2. Select supplier, 3. Contract agreement, 4. Ordering, 5. Expediting, 6. Evaluation, cf. van Weele (2010), pp. 28-31.

meeting. Since there is no contact between the contract parties yet, the LSP is still in the phase of acquisition of customer relationship or customer care.

Phase 1: Bidder search and RFI (Request for Information). As an IC, the first active action to establish relationships on the bidder market is to create and send an RFI. This query to potential candidates by e-mail or telephone should evaluate how much interest, capacity, and capability is available if a potential LSP would participate in the actual tendering process. A closely related prerequisite for further information exchange is to send a Non-Disclosure Agreement (NDA) at the same time. This is an agreement which both contract parties have to sign to protect business-sensitive information from disclosure by either contract party in the scope of the tendering process[217].

Phase 2: Preparation of RFPs and offer. The next phase includes the conception and the actual creation of the tender documents for the tender object. The so-called "Request for Quotation" (RFQ), sometimes also combined with a price query "Request for Pricing" (RFP), contains all the necessary targets, requirements, and dates for the tender object. On this base, and through constant, sometimes time-delayed, exchange, the LSP creates the offer and initiates his own internal project management process[218].

Phase 3: Offer normalization / adjustment and negotiation. After the offers have been received, the IC will carry out a normalization process, which compares, in addition to the verification of the completeness of the offers, the submitted concepts, and prices. The goal is to derive from the previous "longlist" of all providers a "shortlist" of two to three LSPs suitable for the whole project. The IC also carries out an information and communication process, including on-site visits and answering questions, that aims to adjust content and price of the LSP's offer[219].

Phase 4: Offer presentation and choice of partner. The outlined process of offer adjustment is accompanied by one or more offer presentations and feedbacks that can lead to new rounds of negotiations and offer adjustment processes. During the process, a final LSP is gradually emerging from the shortlist. With this partner, a Letter of Intent (LOI) is often drawn up, which shows both parties' intention to establish a business relationship[220].

Phase 5: Contract drafting, negotiation and closing. The most important phase of the tendering process on the IC side is the contract drafting and on the LSP side the contract evaluation. Since contract terms of three or more years are common in a typical contract logistics project, when the LSP provides the offer, he will rely on the communication regarding cost-, time- and risk-driving contract components early on. For example, in Germany, the request for transfer of business according to §613a BGB (Bürgerliches Gesetzbuch, German civil code). However, this means the contract components are communicated and negotiated already in the early stages. Phase 5 runs partly parallel to the other tendering phases. On the way to closing the contract, the LOI plays an important role. This pre-contractual document, in the sense of a "statement of intention to conclude a contract," should ensure that the selected LSP can start with its tasks for the ramp-up phase - such as the final contracting of its suppliers - on time in order to avoid endangering the commonly agreed "go-live date[221]."

Throughout these phases, two teams cooperate until successful contract closure. On the IC side, the "Buying center" consists mainly of members from logistics and purchasing departments; on

---

[217] Pfohl (2016), pp. 345-347;
[218] Pfohl (2016), pp. 345-347;
[219] Pfohl (2016), pp. 345-347;
[220] Pfohl (2016), pp. 345-347;
[221] Pfohl (2016), pp. 345-347;

the LSP side, the "Selling center" mainly consists of members of the operations and sales departments[222]. The six phases are summarized in Figure 4.

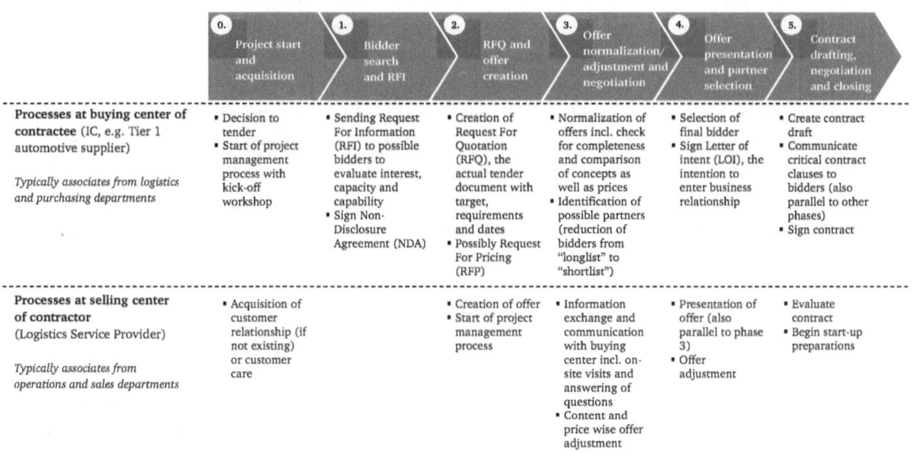

	0. Project start and acquisition	1. Bidder search and RFI	2. RFQ and offer creation	3. Offer normalization/ adjustment and negotiation	4. Offer presentation and partner selection	5. Contract drafting, negotiation and closing
Processes at buying center of contractee (IC, e.g. Tier 1 automotive supplier) *Typically associates from logistics and purchasing departments*	• Decision to tender • Start of project management process with kick-off workshop	• Sending Request For Information (RFI) to possible bidders to evaluate interest, capacity and capability • Sign Non-Disclosure Agreement (NDA)	• Creation of Request For Quotation (RFQ), the actual tender document with target, requirements and dates • Possibly Request For Pricing (RFP)	• Normalization of offers incl. check for completeness and comparison of concepts as well as prices • Identification of possible partners (reduction of bidders from "longlist" to "shortlist")	• Selection of final bidder • Sign Letter of intent (LOI), the intention to enter business relationship	• Create contract draft • Communicate critical contract clauses to bidders (also parallel to other phases) • Sign contract
Processes at selling center of contractor (Logistics Service Provider) *Typically associates from operations and sales departments*	• Acquisition of customer relationship (if not existing) or customer care		• Creation of offer • Start of project management process	• Information exchange and communication with buying center incl. on-site visits and answering of questions • Content and price wise offer adjustment	• Presentation of offer (also parallel to phase 3) • Offer adjustment	• Evaluate contract • Begin start-up preparations

Figure 4: Typical contract logistics tender phases.

Source: Own figure after Pfohl[223].

3.6. Buyer-Supplier relationships

Central to unlocking the benefits of outsourcing is the effective management of the buyer-supplier relationship[224]. This section will provide an overview of various dimensions of BSRs, their benefits and costs, as well as the different types that can be distinguished.

Dimensions

BSRs can vary across a variety of dimensions[225]. *Commitment* refers to an implicit or explicit pledge of relational continuity between exchange partners, typically implying solidarity and cohesion[226]. *Trust* is "the willingness to rely on an exchange partner in whom one has confidence[227]," the belief that a party's word is reliable, and that a party will fulfill its obligations in an exchange relationship[228]. *Cooperation* is the degree to which focal activities to the relationship are carried out jointly, i.e., actions taken by firms in interdependent relationships to achieve mutual outcomes with expected reciprocation over time[229]. *Mutual goals* refer to the degree to which partners share goals that can only be accomplished through common action and the maintenance of the relationship[230]. *Interdependence*, the existence of bilateral dependencies in a dyad, the sum of one actor's dependence on the other actor, and the other

---

[222] Pfohl (2016), pp. 345-347; Anderson et al. (2011), pp. 97-115.
[223] Pfohl (2016), pp. 345-346.
[224] Saeed et al. (2005), pp. 365-390; Tangpong et al. (2008), pp. 571-579.
[225] Wilson (1995), pp. 335-337.
[226] Dwyer et al. (1987), p. 19.
[227] Schurr/Ozanne (1985), p. 940; Moorman et al. (1992), p. 315.
[228] Schurr/Ozanne (1985), p. 940; Moorman et al. (1992), p. 315.
[229] Bensaou (1997), pp. 107-124; Anderson/Narus (1990), p. 45.
[230] Wilson (1995), pp. 335-337.

actor's dependence on the one actor[231]. *Power,* one party's ability to get another party to undertake an activity that the latter party would typically not do[232]. The closely related power imbalance denotes the difference between two actors' dependencies, the ratio of the more powerful actor's power to that of the less powerful actor[233]. *Performance satisfaction* is the degree to which the business transaction meets the business performance expectations of the partner[234]. *Structural bonds* are ties that bind parties together on an organizational level[235]. The interaction of various non-reusable investments dedicated to a relationship increases the costs of exiting and therefore creates a force that ties partners stronger together[236]. The *comparison level of the alternatives* is a standard representing the quality of outcomes a company can expect from a given kind of relationship, based upon present and past experience with similar relationships, and knowledge of other company's similar relationships[237]. *Adaptation* is altering one or both party's products, production process, stockholding, etc. to bring about either initial fit between each other's needs and capabilities or to react to changing business conditions[238]. *Nonretrievable investments* are investments that may not be recovered if the relationship ends[239]. Investments can become nonretrievable, or "transaction specific," if they are site-specific, e.g., located in close proximity to economize on inventory and transportation expenses, physical asset specific, e.g., when specialized base materials are required to produce a component, or human asset specific, for instance knowledge or training[240]. *Shared technology* is the extent to which partners value the technology contributed by the relationship[241]. This variable can refer to technology acquisition from suppliers or EDI (Electronic Data Interchange) adoption[242]. *Social bonds* hold buyers and sellers closely together in a personal sense, including personal interactivity and feelings of personal closeness, such as feelings of likeness, acceptance, or friendship[243]. Based on expected benefits, these variables can be configured and managed in a way that leads to either a rather market-based, arms-length, or a long-term committed, partnership type of BSR[244].

Benefits and costs

Various studies examined the link between effective BSRs and performance benefits from a buyer and supplier side[245]. Buyers can profit from a successful relationship financially and improve innovation and quality while increasing lead time performance and customer responsiveness[246]. Suppliers can also benefit financially, expect future relationship prospects

---

[231] Casciaro/Piskorski (2005), pp. 169-171; Bacharach/Lawler (1981), pp. 220-222.

[232] Anderson/Weitz (1989), p. 312.

[233] Lawler/Yoon (1996), pp. 91-93.

[234] Wilson (1995), pp. 338-339.

[235] Smith (1998), p. 79; Other scholars refer to these types bonds as "Functional bonds" or "task bonds", defined as "the multiplicity of economic, performance, or instrumental ties or linkages that serve to promote continuity in a relationship" Smith (1998), p. 79.

[236] Wilson/Jantrania (1994), p. 57.

[237] Anderson/Narus (1984), p. 63.

[238] Hallen et al. (1991), pp. 30-32.

[239] Wilson/Jantrania (1994), p. 57.

[240] Williamson (1981), p. 555.

[241] Wilson (1995), p. 339.

[242] Hartley et al. (1997), pp. 65-67; Sriram/Banerjee (1994), pp. 30-39.

[243] Gounaris (2005), pp. 129-136.

[244] Ellram/Murfield (2019), pp. 40-41; Duffy/Fearne (2004), pp. 57-59.

[245] Kannan/Choon Tan (2006), pp. 757-763; Terpend et al. (2008), p. 40. Challenges and good practices in measuring the BSR and performance link, i.e. distinguishing outcomes of the relationship itself compared to broader firm level outcomes, are discussed in Kannan/Choon Tan (2006), p. 762.

[246] Carr/Pearson (1999), pp. 514-517; Larson/Kulchitsky (2000), pp. 30-36; Martin/Grbac (2003), pp. 30-36; Corsten/Felde (2005), pp. 454-457; Johnston et al. (2004), pp. 31-37.

(e.g. future sales), improve product and process design, increase their quality, and reduce inventory costs plus lead time[247].

Close BSR can lead to an overall supply chain performance increase in terms of market share, return on assets, product price, product quality, and customer satisfaction levels[248]. Based on a study in the automotive industry, Benton and Maloni showed that an improved BSR enables suppliers to improve their performance by providing better designed, higher quality products at reduced costs, which translates to better cars at competitive prices for the OEM, leading ultimately to enhanced sales, market share, and customer loyalty[249]. Monczka et al. demonstrated a link of characteristics of close BSR and price reductions, quality improvements, new product development time reductions, cycle time reduction, and technology improvements[250]. In a study in the food industry, Stank et al. validated that BSRs can lead to decreased inventory levels, decreased order cycle times, and decreased order cycle variance. They also showed a positive impact on product availability, customer satisfaction, as well as flexibility in meeting customer requirements and assessing customer needs[251]. Dyer analyzed automotive transaction relationships in the USA and Japan and found that BSR as an efficient governance mechanism can simultaneously lower transaction costs and increase relation-specific investments, thereby creating a competitive advantage[252]. According to Heide and John, who conducted studies among industrial firms and their suppliers, BSRs emerge as responses to the need to safeguard transaction-specific assets and adapting to uncertainty. Close BSRs then lead to increased joint action, such as technology transfer in the form of joint component testing/prototyping[253].

However, while BSRs can lead to improved performance, they also come at a cost. Gadde and Snehota proposed four types of costs that occur through BSRs. *Direct procurement* costs are "what shows up as the invoice from the supplier," *direct transaction costs* are "costs of transportation, goods handling [and] ordering," *relationship handling costs,* that is, "continuous interaction costs for maintaining the relationship," and *supply handling costs,* "costs for the purchasing organization as a whole, including communication and administrative systems, warehousing operations, [or] process adaptations[254]."

Types

To account for the benefits of the BSRs while balancing their cost, companies should employ a differentiated approach[255].

Drawing from his research in the automotive industry, Bensaou proposes a portfolio matrix based on low/high buyer-specific investments and low/high supplier-specific investments. He distinguishes the contextual factors product, market, and supplier characteristics, and developed management profiles for each of the four resulting contextual profiles[256]. For example, the profile with high buyer's specific investments and high supplier's specific investments is designated *Strategic Partnership.* In this profile, products require a high level of customization, are technically complex (or an integrated subsystem) and based on new

[247] Kalwani/Narayandas (1995), pp. 8-15; Kotabe et al. (2003), pp. 304-312; Duffy/Fearne (2004), pp. 61-67.
[248] Narasimhan/Nair (2005), pp. 307-312.
[249] Benton/Maloni (2005), pp. 9-19.
[250] Monczka et al. (1998), pp. 563-570.
[251] Stank (1999), pp. 31-37.
[252] Dyer (1997), pp. 540-553.
[253] Heide/John (1990), pp. 32-35.
[254] Gadde/Snehota (2000), pp. 306-316.
[255] Bensaou (1999), pp. 37-43.
[256] Bensaou (1999), pp. 36-44.

technology which is close to the buyer's core competency. Tight mutual adjustments are needed in key processes, design changes are frequent, and innovation leaps occur in technology, product, or process. This demands strong engineering expertise and large capital investments. The market is coined by strong demand and high growth, fierce competition, and high concentration, as well as frequent changes in competitors due to unstable or lack of dominant design. Buyers maintain design and testing capability in-house. The suppliers (in this profile termed "partners") are large multiproduct supply houses with strong proprietary technology that are active in research and innovation and possess strong skills and capabilities in design, engineering, and manufacturing. The corresponding management profile suggests that information is shared frequently, in "broadband," and there is a "rich media" exchange with regular mutual visits, and the practice of guest engineers is employed. Boundary spanners' tasks are ill-defined, ill-structured, nonroutine with frequent unexpected events, and a large amount of time is spent with suppliers' staff, mostly on coordinating issues. The relationship is characterized by high mutual trust and relationship commitment, with a strong sense of buyer fairness. Suppliers are early involved in the design process, and extensive joint action and cooperation occur. Suppliers have an excellent reputation[257].

Olsen and Ellram proposed a different profile approach. They recommend a three-step procedure to analyze a company's supplier relationships and develop corresponding actions[258]. In step 1, the company's purchases should be analyzed based on the difficulty of managing the purchasing situation and the strategic importance of the purchase. In step 2, the supplier relationships must be analyzed. Here, a portfolio model to categorize the supplier relationships is developed based on the relative supplier attractiveness and the strength of the relationship[259]. In step 3, by comparing the analyses conducted in steps 1 and 2, action plans should be developed. Supplier relationships could be strengthened, the supplier attractiveness improved, or resources reallocated that are currently assigned to a relationship.

Another classification for supplier relationships is based on different levels of commercial and technical complexity[260]. Recommendations are also proposed based on a supplier's position in a corresponding portfolio matrix. For instance, suppliers with a high degree of technical complexity and a high degree of commercial complexity require *Supplier Management*, that means they require collaboration and coordination of supply chain activities, centralized management (control), and high-level involvement with a time horizon of up to 10 years, so that strategic alliances and long-term strategies should be used. On the other side, suppliers with a low degree of technical complexity and a low degree of commercial complexity require *Purchasing Management,* which is coined by procurement efficiency for commodity items, relies on standard product specifications and order processes to minimize transaction costs, and short planning horizons of typically less than 12 months. Here, information is shared only if necessary, to facilitate transaction[261].

---

[257] Bensaou (1999), pp. 36-44.

[258] Olsen/Ellram (1997), pp. 103-111.

[259] Factors influencing the *relative supplier attractiveness* are financial and economic factors (e.g. the supplier's margins or financial stability), performance factors (delivery, quality, price), technological factors (e.g. the supplier's design capabilities or speed in development), organizational, cultural, and strategic factors (e.g. the strategic fit between buyer and supplier or the top management capability) as well as other factors (e.g. the ability to cope with changes in the environment). Factors that describe the strength of the relationship are economic factors (e.g. the volume of purchases or the importance of the buyer to the supplier), the character of the exchange relationship (e.g. types of exchange, the number of other partners), the cooperation between buyer and supplier (e.g. cooperation in development, technical cooperation) and the distance between the buyer and the supplier (social, cultural, technological, time and geographic distance). Olsen/Ellram (1997), pp. 103-111.

[260] Spekman et al. (1999), pp. 113-115.

[261] Spekman et al. (1999), pp. 113-115.

However, Gadde and Snehota argue that such portfolio approaches and recommendations are an oversimplification which might not lead to the best use of BSRs[262]. In practice, many cooperations fail expectations because little attention is given to interpersonal connections between boundary spanners who work closely together and serve to shape the cooperation[263].

3.7. Boundary spanners

A minimal defining characteristic of formal organizations are boundaries that protect the members of the system from extra-systemic influences, and that regulate the flow of information, material, and people into or out of the system that surrounds it[264]. People who operate at an organization's boundary and manage the interface between organizations and their environments as "linking pins" or "gatekeepers" are called boundary spanners[265]. They perform two types of functions, information processing, and external representation or gatekeeping [266]. In their *information processing function*, they search, filter, summarize information, and facilitate information dissemination[267]. Therefore, the boundary spanner must be well connected internally and externally, be able to translate across communication boundaries, and be aware of contextual information inside and outside of their organizational boundaries[268]. Under the *external representation* or gatekeeping function fall roles that involve resource acquisition and disposal, political legitimacy and hegemony, social legitimacy, and organizational image. This includes departments related to a) resource acquisition and disposal, such as purchasing, marketing, and sales representatives or personnel recruiters, b) mediation among groups, e.g., corporate lawyers, or c) increasing of an organization's visibility, for instance, public relations[269].

Working in various roles simultaneously generates various tensions and paradoxes, including working with autonomy and interdependence, being participative and authoritarian, balancing advocacy and inquiry, and being able to manage conflict using effective bargaining and negotiation skills[270]. People working in a boundary function, such as a salesperson, are at high risk of job burnout[271]. This is due to various factors such as role ambiguity, the degree of uncertainty about the job function and responsibilities, role conflicts, perceptions of incompatible demands and expectations about the functions and responsibilities, and a lack of support within and outside the organization[272]. Moreover, it can be due to work overload, the perception that one has too many tasks to finish in a given time, depersonalization, the display of negative attitudes to customers or coworkers caused by a lack of control over key aspects of the job, and personal non-accomplishment, the feeling that even the best efforts are not producing intended results or are not being recognized[273].

---

[262] Gadde/Snehota (2000), pp. 306-316.

[263] Hutt et al. (2000), pp. 51-62.

[264] Leifer/Delbecq (1978), p. 41; Thompson (1962), pp. 309-324; Aldrich/Herker (1977), pp. 217-221.

[265] Organ (1971), p. 73; Leifer/Delbecq (1978), pp. 40-41; Katz/Kahn (1978), 17-743; Utterback (1971), p. 84. An overview about further terms used for Boundary spanning individuals can be found in Leifer/Delbecq (1978), p. 42.

[266] Adams (1976), pp. 1175-1199.

[267] Aldrich/Herker (1977), pp. 218-221.

[268] Tushman/Scanlan (1981), pp. 300-303.

[269] Aldrich/Herker (1977), pp. 218-221.

[270] O'Leary/Bingham (2009), pp. 255-269.

[271] Hollet-Haudebert et al. (2011), pp. 413-424; Greenglass et al. (2003), pp. 581-594; Rizzo et al. (1970), pp. 155-162; Lewin/Sager (2008), pp. 240-243.

[272] Hollet-Haudebert et al. (2011), pp. 413-424; Greenglass et al. (2003), pp. 581-594; Rizzo et al. (1970), pp. 155-162; Lewin/Sager (2008), pp. 240-243.

[273] Hollet-Haudebert et al. (2011), pp. 413-424; Greenglass et al. (2003), pp. 581-594; Rizzo et al. (1970), pp. 155-162; Lewin/Sager (2008), pp. 240-243.

Boundary spanning is highly complex, as boundary spanners must act simultaneously as reticulists, interpreter and communicator, coordinators as well as entrepreneurs and innovators[274]. They, therefore, have to possess a variety of skills. In their *reticulist* role, they must be able to develop and sustain a network of interpersonal relationships, influence actors and agencies with differential power bases, perform at both strategic and operational levels via network management techniques and build coalitions and support around issues and strategies through effective communication, negotiation, and political skills[275]. As *interpreter and communicator*, they have to initiate and sustain effective interpersonal relationships built upon skills such as trust, communication, listening, empathy, negotiation, diplomacy, and conflict resolution[276]. For this, they also have to be literate in the "professional languages" of the various partnership domains where they have to act as a broker[277]. In the *coordinator* role, boundary spanners must organize, plan, coordinate, and service the "collaborative machinery," which involves knowledge management, i.e., communicating and sharing information equally and transparently, with various actors[278]. Furthermore, in the role as *entrepreneur and innovator*, boundary spanners must be flexible, creative, and laterally thinking, combining resources in response to opening windows[279]. They have to be risk-taking, self-disciplined and perseverant, action and goal-oriented and must have a desire to succeed[280].

Challenging for the management of buyer-supplier relationships and boundary spanners' involvement is that they operate under a company's name. They represent their employer, which in return, must trust these employees to fulfill their role conscientiously. A high degree of freedom for them can lead to improved outcomes, but also allows for opportunistic behavior that can potentially damage a company[281].

---

[274] Williams (2013), pp. 24-28.
[275] Williams (2011), pp. 27-30; Eckhardt/Rosenkranz (2010), pp. 75-82.
[276] Millman/Wilson (1995), pp. 13-18; Vangen/Huxham (2003), pp. 10-27; Williams (2011), pp. 27-30.
[277] Trevillion (1992), 103-109.
[278] Williams (2011), pp. 27-30.
[279] Kingdon (2014), pp. 165-208; Williams (2002), p. 110.
[280] Gartner (1988), pp. 12-27.
[281] Perrone et al. (2003), pp. 434-435; Luo (2006a), pp. 141-144.

4. Guanxi

The theme of this chapter is Guanxi. As illustrated in Figure 5, this is the second key concept that lays the foundation for the subsequent explorative study.

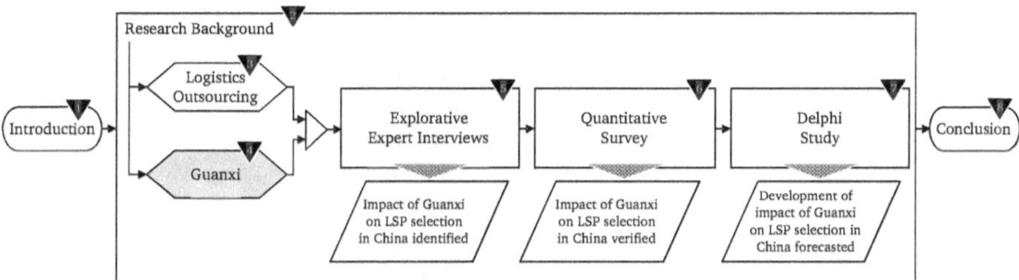

Figure 5: Chapter "Guanxi" within overall research process
Source: Own figure

In section 4.1, various definition approaches for the phenomenon will be introduced, and the meaning behind the Chinese characters that the phrase consists of explained. As the concept is multifaceted and defies an exact definition, it is vital to understand the nature of Guanxi. Therefore, in section 4.2, the term's underlying dimensions ganqing, renqing and xinren are carved out. In addition, in section 4.3, various typologies of Guanxi ties are introduced. Guanxi can also be extended to the organizational level as "aggregated Guanxi." This sum of all employees' social networks and ties can lead to various benefits and costs with other companies and government officials, as will be described throughout sections 4.4 and 4.5. Subsequently, the closely related debate about the relationship between Guanxi and corruption is discussed in section 4.6. Equally controversial is the question whether Guanxi is unique to China, exists in some other cultures, or is even universally practiced everywhere. These perspectives will be, therefore, illuminated in the closing section 4.7 of this chapter.

4.1. Definition

Guanxi is considered a multifaceted, cultural phenomenon that defies an exact definition[282]. Also, no English expression has exactly the same meaning, and no precise translation exists, so that, for instance, popular English glossaries like the Oxford English Dictionary or the Collins English Dictionary directly include the term[283]. To comprehend the word, it is helpful to understand the meaning behind the two Chinese characters that it consists of[284]. "Guan" (in Chinese characters "关") as a noun can be translated as a pass, barrier, gate, or hurdle[285]. As a verb, it signifies to relate, to receive, to be concerned, or to close[286]. "Xi" ("系") as a noun means a system, tie, relationship, or connection[287]. Translated as a verb, it stands for to tie up, to link,

---

[282] Fan (2002b), p. 546.
[283] Dunfee/Warren (2001), p. 192; Oxford English Dictionary (2017b); Collins English Dictionary (2017).
[284] Lee/Dawes (2005), p. 29.
[285] Fan (2002b), p. 546; Lee/Dawes (2005), p. 29.
[286] He et al. (2015) quoted from an earlier edition after Langenberg (2007), p. 5.
[287] Fan (2002b), p. 546; He et al. (2015) quoted from an earlier edition after Langenberg (2007), p. 5; Lee/Dawes (2005), p. 29.

or to care for[288]. Lee and Dawes, therefore, conclude that Guanxi literally means "pass the gate and get connected[289]." However, considering the diverse meaning of the individual characters that the term is composed of, and taking into account that it is a social phenomenon affected by the "fluctuating stream of history" which is consistently adapting, no uniform definition of Guanxi is universally accepted[290]. Common definition attempts revolve around "relationship," "special relationship," "connection," or "particularistic ties[291]."

Some define Guanxi by describing the characteristics of the relationship. Here, undisputed themes are that Guanxi is personal and reciprocal[292]. Chen et al. also highlight that Guanxi is rather informal than formal, Davison et al. add that it is strong, binding, and long-lasting, and Alston emphasizes that it involves being fully committed[293].

In a similar direction goes the definition based on the purpose of the relationship. According to Hackley and Dong, Guanxi is a "special kind of social connection, linking two individuals to enable a social interaction and exchange[294]." Furthermore, Pearce and Robinson see Guanxi as a "network of relationships.[295]" They consider the purpose of the relationship as an essential component, pointing out that the network is cultivated "to attain mutual benefits[296]." This notion is also shared by Luo, who explains that Guanxi is an "intricate and pervasive relational network which Chinese cultivate energetically, subtly, and imaginatively[297]." It contains "implicit mutual obligation, assurance and understanding, and governs Chinese attitudes toward long-term social and business relationships[298]."

Other scholars define Guanxi as a distinctive form of social capital in Chinese societies[299]. Gold et al. support this view by arguing that Guanxi is clearly accumulated with the intention to convert it into economic, political, or symbolic capital[300]. Guanxi is also considered as a synonym for "special favors and obligations to the Guanxi circle[301]."

In contrast, Ai recognizes that people use the term Guanxi to refer to someone who knows plenty of people, is well connected, and gets things done - not necessarily through formal channels[302]. Dong et al. discovered in their fieldwork that when individuals refer to "having (good) Guanxi," they mean that they have a good relationship or regular contact with somebody of importance[303].

A different view is taken by Fan, who sees Guanxi as a multi-path process that starts with two parties and moves on, involving more parties, and will stop only when a solution is finally found, or the task is abandoned[304]. Guanxi, therefore, refers to the "establishment of a connection between two independent individuals to enable a bilateral flow of personal or social

---

[288] Fan (2002b), p. 546; He et al. (2015) quoted from an earlier edition after Langenberg (2007), p. 5.

[289] Lee/Dawes (2005), p. 29.

[290] Langenberg (2007), p. 5; Yang (2002), p. 459.

[291] Alston (1989), p. 28; Yeung/Tung (1996), p. 55; Ai (2006), p. 105; Jacobs (1979), p. 238.

[292] Xin/Pearce (1996), p. 1642; Björkman/Kock (1995), p. 532; Lee/Dawes (2005), p. 29; Pye (1982), p. 89; Hackley/Dong (2001), p. 17; Wu/Chiu (2016), p. 3399.

[293] Chen, Chao C. et al. (2004), p. 200; Davison et al. (2017), p. 225; Alston (1989), p. 28.

[294] Hackley/Dong (2001), p. 17.

[295] Pearce/Robinson (2000), p. 31.

[296] Pearce/Robinson (2000), p. 31.

[297] Luo (1997), p. 44.

[298] Luo (1997), p. 44.

[299] Chen, Ming-Huei et al. (2015), p. 900.

[300] Gold et al. (2002), p. 7.

[301] Lee/Dawes (2005), p. 29.

[302] Ai (2006), p. 105.

[303] Dong et al. (2017), p. 334.

[304] Fan (2002b), p. 551.

transactions [where] both parties must derive benefits from the transaction to ensure the continuation of such a relationship[305]."

Considering these various elements of the term, Bian makes a multifaced definition approach. Guanxi can then be "the existence of a relationship between people who share a status group or are related to a common person", "actual connections or contact between people. How often they are in contact, how well they know each other, or how much they like each other reflects their degree of Guanxi," or "people with whom one has a strong connection[306]."

4.2. Dimensions

Although such definition approaches are useful, to truly clarify the concept, it is vital to understand the nature of Guanxi[307]. Hence, in this section, the term's underlying dimensions ganqing, renqing and xinren will be carved out.

Ganqing

Ganqing consists of the two characters "Gan" (Chinese "感") that can be translated as one's feeling and "Qing" (Chinese "情") that can be translated as affection, sentiment, or emotion[308]. Ganqing is, therefore, often translated as feelings or affection[309]. It is the emotional side of Guanxi and refers to the "degree of emotional understanding and connection," "emotional attachment among members of networks," or "positive mutual feelings to one another[310]." It is related to enduring, emotional commitments found in long-term, intimate social bonds and signifies a sense of loyalty or good rapport between two actors[311]. Ganqing is one of the indicators to describe the quality of a relationship between two parties – "the better the ganqing, the closer the Guanxi[312]." Also, the Chinese expression "you ganqing" (Chinese "有感情"), which means "to have ganqing," implies the prevalence of feelings and a bond between two parties[313]. When ganqing is developed, a business partner is perceived more like a friend than a business acquaintance[314]. Ganqing is cultivated through social interaction and continuous exchange of mutual help[315]. By sharing positive and negative feelings and participating in mutually enjoyed social activities such as wining and dining, joining events, or toasting and serving food at banquets, ganqing can grow[316]. The underlying understanding of a ganqing relationship is that everybody must assist their friends if they encounter difficulties[317]. If there is this kind of relationship between two people or parties, they are willing to take care of each other under all circumstances, can press for advantages from the other, and may expect special

---

[305] Yeung/Tung (1996), p. 55.
[306] Bian (1994), p. 96.
[307] Fan (2002b), p. 546.
[308] Yen et al. (2017), pp. 105-107.
[309] Barnes et al. (2011), pp. 511-512.
[310] Chen/Chen (2004), p. 314; Barnes et al. (2011), pp. 511-512; Wang (2007), p. 82; Berger et al. (2017), p. 2.
[311] Lee/Dawes (2005), p. 35; Chen/Chen (2004), p. 314; Berger et al. (2017), p. 2.
[312] Kipnis (1997), p. 157.
[313] Yen et al. (2011), p. 99.
[314] Yen et al. (2017), pp. 105-107.
[315] Mavondo/Rodrigo (2001), p. 113; Barnes et al. (2011), pp. 511-512.
[316] Chen/Chen (2004), p. 314; Yen et al. (2011), p. 99; Barnes et al. (2011), pp. 511-512; Kipnis (1997), p. 27; Some scholars therefore refer to Ganqing even as "bonding," cf. Lee et al. (2018), pp. 357-358.
[317] Luk et al. (1999), p. 260.

consideration[318]. Such an established emotional bond can also help to reduce hatred or lower the probability of having emotional, direct confrontations[319].

Renqing

Another dimension of Guanxi is renqing, which is composed of "ren," the character for human being (Chinese "人") and "qing," the character for affection, sentiment, and emotion" (Chinese "情")[320]. Renqing can have a variety of meanings such as empathy, an emotional responses of an individual confronting various situations of daily life, a resource that an individual can present to another as a gift within social exchange, or "a set of social norms by which one has to abide in order to get along well with other people[321]." Hence, renqing refers to an informal social obligation to another party that results out of the engagement in a Guanxi relationship[322]. It is a cultural value that underpins the exchange of favors and gifts among friends, but also plays the role of generating a sense of mutual dependence, obligation, and indebtedness[323]. Renqing involves exchanges of both emotional and economic favors according to certain social norms and is like owing of a favor[324]. Elements of renqing are reciprocity, the anticipation of repayment, and empathy[325]. If reciprocity is not carefully observed, the other party will not only be unable to function in the former relationship, but an incumbent might also lose opportunities for future interactions with the Guanxi network[326]. This obligation comes from intense social pressure and individual psychological needs[327]. With the exchange of favors, the other party is, therefore, "taken hostage" to mutual commitment[328]. Renqing is built through intangible, such as advice, information and counseling, or tangible favors, like gifts, jobs, houses, or other products and services[329]. Gifts exchanged among two parties in a Guanxi relationship can range from small personal presents such as cards, clothes, art objects, and theatre tickets to business-related deals like exclusive representation agreements, efficient service, or favorable credit terms[330]. The greater the exchange of favors, the closer the two parties[331]. Overall, renqing provides the moral foundation for the ideals of "reciprocity and equity," which are part of every Guanxi relationship[332].

Xinren

Xinren, which can be translated as trust or deep trust, is the cognitive component of Guanxi[333]. The term is composed of two characters. The first one, "xin" (Chinese "信"), a verb that means to trust and to believe in, is also used in the word for confidence (Chinese "xinxin", "信心") and

---

[318] Chen/Chen (2004), p. 314; Luk et al. (1999), p. 260.

[319] Yen et al. (2017), pp. 105-107.

[320] Shi et al. (2011), p. 497; Yen et al. (2017), pp. 105-107.

[321] Hwang (1987), pp. 953-954; Shou et al. (2011), p. 504.

[322] Luo (1997), p. 45.

[323] Lin (2002), pp. 59-60; Lisha et al. (2017), pp. 1131-1132; Berger et al. (2015), p. 168; Wang et al. (2008), p. 820; Lee et al. (2018), pp. 357-358.

[324] Yang/Wang (2011), p. 493; Li et al. (2016), p. 20.

[325] Hwang (1987), p. 957; Wang et al. (2014), p. 663; Wang et al. (2008), p. 820.

[326] Leung et al. (2011), pp. 1197-1198; Wang et al. (2008), p. 820.

[327] Wang et al. (2014), p. 663.

[328] Choi et al. (2011), pp. 59-60; Berger et al. (2015), p. 168.

[329] Fan (2002b), p. 549.

[330] Yen et al. (2011), p. 105.

[331] Barnes et al. (2011), p. 511; Yen et al. (2011), p. 100.

[332] Luo (1997), p. 45.

[333] Chen/Chen (2004), pp. 313-314; Kriz/Keating (2010), p. 308; Berger et al. (2015), pp. 168-169.

belief (Chinese "xinnian", "信念"), thus emphasizing sincerity[334]. The second character, "ren" (Chinese "任"), stands for "heavy responsibilities that one can carry[335]." The colloquial understanding of xinren is that one can believe in another's word[336]. Within the social norms of Guanxi, xinren refers to trustworthy persons, meaning they are sincere, honest, reliable and capable[337]. This definition includes the elements sincerity, or in the Western distinction benevolence, and usability, or employability, which is based on one's ability[338]. Chen et al. argue that the sincerity-based element might be more critical, as Guanxi always focuses on individuals and benevolence relates to a "person as a whole," while one's ability might be domain-specific[339]. Xinren is founded on an individual's history, reputation, and experience in previous transactions[340]. Kriz and Keating showed in an explorative study in China that to create trust, it is important to keep promises ("whatever the reason you must keep your word"), ensure mutual honesty, and that certain feelings must be present ("something in my heart")[341]. When the unspoken obligations of a Guanxi relation are followed, a participant in the Guanxi network owns deep trust from other network participants. However, if the unspoken obligations are not followed, face[342] and trust are lost, and related parties' feelings hurt[343]. Giving and receiving help is considered another vital construct of xinren - "you will offer me help. And I would do exactly the same, and that's how you build up trust[344]". Trust decreases defensiveness mechanisms and opportunistic behavior, thus facilitating efficient transactions, which reduces risks[345]. Individuals in a trusting relationship are more willing to contribute ideas, share information, explain goals, clarify problems, and generally approach the relationship with a problem-solving orientation[346]. Suggestions in China to test trust include looking at a person's eyes as well as getting a person drunk and listening to their "loose tongue[347]."

4.3. Typologies

Various typologies were developed to classify the nature of Guanxi ties[348].
The quality of the interaction between individuals is affected by the existence and type of Guanxi bases[349]. These Guanxi bases depend on a specific shared common identification, either preordained or voluntary[350]. *Preordained bases* are blood-based, include being own family members, relatives, or members of the same clan[351]. *Voluntary bases* can be either due to a common social identity, a common third party, or anticipatory[352]. Common social identity refers to the joint identification with a hometown, school, workplace (military and civilian) or political

---

[334] Yen et al. (2017), pp. 105-107; Kriz/Keating (2010), pp. 308-309.
[335] Yen et al. (2017), pp. 105-107.
[336] Kriz/Keating (2010), pp. 308-309.
[337] Mayer et al. (1995), p. 715.
[338] Mayer et al. (1995), p. 715; Chen/Chen (2004), pp. 313-314.
[339] Chen/Chen (2004), pp. 313-314.
[340] Berger et al. (2015), pp. 168-169.
[341] Kriz/Keating (2010), pp. 308-309.
[342] "Mianzi", cf. section 2.3.
[343] Park/Luo (2001), pp. 456-459.
[344] Kriz/Keating (2010), pp. 308-309.
[345] Grover et al. (2002), pp. 233-236; Wong/Leung (2001), pp. 72-73.
[346] Sheu et al. (2006), pp. 25-46; Barnes et al. (2011), p. 512.
[347] Kriz/Keating (2010), pp. 308-309.
[348] Chen et al. (2013), pp. 170-171.
[349] Chen et al. (2013), p. 170.
[350] King (1991), pp. 65-70; Jacobs (1979), p. 243.
[351] Tsang (1998), p. 65.
[352] Chen/Chen (2004), pp. 311-312.

party[353]. A Guanxi base for individuals who do not have a direct base can be a joint third party, with which these individuals have Guanxi, which acquaints them [354]. Guanxi can also be anticipatory, where individuals base the relationship on a shared promise to engage in future social or business interactions[355]. These Guanxi bases are not mutually exclusive so that, for example, colleagues can still know a third person or intend to work together in future joint projects.[356]

Another typology is tie based, where Guanxi relationships are distinguished in expressive ties, mixed ties, and instrumental ties [357]. These types are commanded by different sets of interpersonal regulations, have different social and psychological meanings to the involved persons, and vary in the degree of strength and closeness[358]. *Expressive ties* are typically prevalent among family people, "jiaren" (Chinese "家人")[359]. These relationships are very permanent and stable with a long-term focus[360]. Expressive ties are emotional and informal, where egalitarian benefits are sought[361]. *Mixed ties* can be found among familiar people, "shuren" (Chinese "熟人") like close friends or coworkers[362]. They are a mixture of expressive and instrumental ties, although the boundaries between mixed ties and instrumental ties are somewhat fluid[363]. Such relationships are moderately permanent and stable with a long-term focus [364]. They are emotional, relatively informal, and mutual benefits are sought [365]. *Instrumental ties* are prevalent with strangers, "shengren" (Chinese "生人"), such as sellers and buyers of goods and services[366]. These relationships are unstable, temporary, and with a short-term focus[367]. Instrumental ties are utilitarian, formal, and arm's length dealings are sought[368]. Contrary to this clear distinction, Yang argues that all types of Guanxi relationships involve the exchange of emotions and mutual benefits, while just varying in the extent of the given type of exchange[369].

A different Guanxi typology, distinguishing Family, Helper, and Business Guanxi, mainly based on underlying values, was developed by Fan[370]. *Family Guanxi* is a special relationship rooted in Chinese cultural values. Core values are affection, obligations, and empathy, and it is mostly blood-based, with some social base. The purpose is emotion-drive and mutually dependent, and it is emotional plus instrumental by nature. Love or affection gets exchanged, and it is seen as a normative obligation where reciprocity is not necessary. This type of Guanxi is solid and stable over an extended timeframe. A downside can be nepotism. *Helper Guanxi* is a process of exchanging favors, rooted in cultural values and contemporary socioeconomic factors. Core values are face, trust, and credibility. It is mostly social based. The motivation is "to get things

---

[353] Tsang (1998), p. 65; Chen et al. (2013), pp. 171-172. These commonalities have to be understood as shared institutions rather than necessarily contemporaneous experiences, cf. Chen/Chen (2004), pp. 311-312.

[354] Chen/Chen (2004), pp. 311-312.

[355] Chen/Chen (2004), pp. 311-312.

[356] Chen/Chen (2004), pp. 311-312.

[357] Hwang (1987), pp. 949-953.

[358] Tsui/Farh (1997), pp. 61-62.

[359] Yang K. S. (1993), pp. 19-56.

[360] Tsui/Farh (1997), pp. 61-62.

[361] Lee et al. (2001), p. 53.

[362] Yang K. S. (1993), pp. 19-56.

[363] Hwang (1987), pp. 952-953.

[364] Lee et al. (2001), p. 53.

[365] Lee et al. (2001), p. 53.

[366] Yang K. S. (1993), pp. 19-56; Chen/Chen (2004), p. 308.

[367] Lee et al. (2001), p. 53.

[368] Lee et al. (2001), p. 53.

[369] Yang (2001), pp. 3-26 quoted after Chen/Chen (2004), p. 309.

[370] Fan (2002b), p. 551.

done", so it is very utility-driven and utilitarian by nature, where mainly favors get exchanged. Reciprocity is expected, although the weaker party benefits more. This Guanxi type is of medium quality, usually unstable, and varies over time or is a one-off. A downside is the burden of indebtedness. *Business Guanxi* is a process of finding business solutions through personal connections rooted in current political and economic structures, i.e., a weak legal system. Core values are face and power or influence. It is mostly based on intermediaries. The purpose is to acquire scarce resources or get special treatment, and it is purely utilitarian by nature. It is a "money and power deal," strictly reciprocal with "gain and loss bargaining." The closeness can vary in this type of Guanxi, depending on the existence of other Guanxi bases, and it is relatively temporary. Downsides are corruption and social loss[371]. Despite proposing this typology, Fan acknowledged that Guanxi is far more complicated in reality and that there might be no clear-cut boundary between these Guanxi types. The typologies are also permeable so that a Guanxi relationship can evolve from one type to another over time[372].

4.4. Benefits

Although Guanxi originates from the relationships of individuals, it can be extended to the organizational level[373]. This "aggregated Guanxi," the sum of all employees' social networks and ties, can have various benefits when applied with other companies or government officials[374]. They will be described in this section.

Guanxi with other companies

Guanxi can play a critical role in gaining valuable and relevant information, especially for which an "atmosphere of trust" is needed[375]. It enhances a managers' ability to acquire useful information and can lead to the obtainment of information about the market, prices, business opportunities, customers, or competitors[376]. Foreign firms can also overcome their "liability of foreignness" through building ties with the local business community, leading to the transfer of knowledge and shared learning[377]. Ties with managers at other firms can also positively impact a firm's market share and enhance its business[378]. A meta-analysis of 53 studies published from 1997 to 2010 by Luo et al. confirmed a positive Guanxi-firm performance link. It showed that business ties have a positive, significant effect on especially non-economic aspects related to an organization's social and societal relationships, such as customer satisfaction or customer loyalty, and competitive success factors such as innovation, productivity, or marketing effectiveness[379]. Due to the way Guanxi deals with governance problems associated with bounded rationality and opportunism, a Guanxi-based exchange might also lead to a transaction cost advantage[380]. Guanxi reduces transaction costs, mainly related to environmental uncertainties such as communication, negotiation, and coordination costs[381]. It is also theorized

---

[371] Fan (2002b), p. 551.

[372] Fan (2002b), p. 551.

[373] Brass et al. also conclude that variables that explain the formation of interpersonal networks also explain the creation of inter-organizational networks. They state that "this is not surprising, since inter-organizational relations are often initially created by boundary spanners," cf. Brass et al. (2004), p. 802; Luo et al. (2012), p. 140.

[374] Geng et al. (2017), pp. 10-11; Peng/Luo (2000), pp. 497-499.

[375] Shin et al. (2007), pp. 171-172.

[376] Chen, Ming-Huei et al. (2015), p. 903; Wiegel/Bamford (2015), p. 315.

[377] Li et al. (2009), pp. 343-349.

[378] Peng/Luo (2000), p. 494.

[379] Venkatraman/Ramanujam (1986), p. 804; Wood (1991), pp. 697-711; Luo et al. (2012), pp. 158-161.

[380] Standifird/Marshall (2000), p. 30.

[381] Davies et al. (1995), pp. 208-213; Lee/Humphreys (2007), pp. 458-464. An informant quoted by Kiong and Kee explains that "For someone new, you may want written contracts. But later, when you get to know him better, there is no need. We may

to prevent opportunistic behavior[382]. As noncompliance in a Guanxi network, which is violating its reciprocal nature, leads to a loss of face and reduces future exchange opportunities, Guanxi acts as a safeguard mechanism from opportunistic behaviors[383].

These benefits also impact supply chain management. Companies can use Guanxi as a strategy for mitigating supply chain risks, especially towards their suppliers in China[384]. The favor exchange nature of Guanxi requests supply chain partners to support each other when in need, which can contribute to risk mitigation[385]. Guanxi can also support strategic purchasing, the "process of planning, implementing, evaluating, and controlling strategic and operating purchasing decisions for directing all activities of the purchasing function toward opportunities consistent with the firm's capabilities to achieve its long-term goals[386]." It includes strategic focus, strategic involvement, and visibility/status of the purchasing function[387]. Firms with a high-level Guanxi corporate culture are more likely to consider key suppliers' relationship as highly important. Therefore, the purchasing function responsible for developing and maintaining this relationship is more likely to be considered as a strategically important function[388]. By creating an alternative, informal, efficient governance system, Guanxi can help to reduce transaction costs and thus facilitate outsourcing arrangements[389]. A firm can increase its competitiveness through a resource concentration on core competencies and the strategic outsourcing of other activities to external companies[390]. It is also supported by the strong emphasis of Guanxi on relationships[391]. As Guanxi is related to the exchange of favors with partners, and supplier development activities can be perceived as a form of favor giving between a buying firm and its suppliers, it further influences supplier development[392]. Supplier development refers to "any set of activities undertaken by a buying firm to identify, measure and improve supplier performance and facilitate the continuous improvement of the overall value of goods and services supplied to the buying company's business unit[393]." To adopt green supply chain management (GSCM) practices, a strong relationship between a buying and supplying firm is crucial, and as Guanxi strengthens such relationships, it can support the introduction of such initiatives[394]. GSCM refers to the integration of environmental thinking into supply chain management, which can include product design, material sourcing, and selection, manufacturing processes, delivery of the final product as well as a product's end-of-life management[395]. These practices can lead to positive environmental and economic performance[396]. A study by Luo et al. could prove this relation, but it also showed that buyer's

---

even give him credit, or cash even, based on [trust]. No need for banks [to guarantee payment] or to sign anything," Kiong/Kee (1998), pp. 88-89.

[382] Standifird/Marshall (2000), p. 30; Gu et al. (2008), p. 17.

[383] Cheng et al. (2012), p. 11; Wang (2007), p. 85.

[384] Abramson/Ai (1999), pp. 21-31; Cheng et al. (2012), p. 11.

[385] Lee/Humphreys (2007), pp. 462-463.

[386] Carr/Smeltzer (1997), p. 201.

[387] Paulraj et al. (2006), p. 108.

[388] Lee/Humphreys (2007), p. 462.

[389] Davies et al. (1995), pp. 208-213; Lee/Humphreys (2007), p. 463.

[390] Quinn/Hilmer (1994), p. 43.

[391] Lee/Humphreys (2007), pp. 454-462.

[392] Lee/Humphreys (2007), pp. 455-462; Lee and Humphreys also investigated how Chinese buyers mitigate the risk for such transaction-specific investments. Buyers focus on suppliers that are proactive in the cumulative development of their own capabilities. By investing in short-term asset specific transactions to improve their performance, suppliers provide justifications for a buyer to support further supplier development activities, cf. Lee/Humphreys (2007), p. 462.

[393] Krause et al. (1998), p. 40.

[394] Simpson et al. (2007), pp. 42-44; Cheng et al. (2012), p. 11.

[395] Srivastava (2007), pp. 54-55.

[396] Zhu/Sarkis (2004), pp. 282-285.

investments in GSCM practices can be insufficient due to the efforts needed to build and maintain Guanxi links[397].

Guanxi with government officials

Guanxi also has certain benefits when applied to government officials, especially as China is often characterized as "government of people" instead of "government of law[398]." Guanxi networks can be a useful means to gain valuable and relevant information on government policies, which is crucial as country regulations are subject to local governments' interpretation[399]. Information that is officially distributed by the government is also recognized to remain at an "uncodified level, leading to potential confusion," so that firms can benefit from personal government relations to decode such information[400]. Ties with government officials further allow companies to obtain information about new regulations or policy changes upfront[401]. Guanxi with government officials can also help to enhance business. It could be demonstrated that ties with government officials significantly affect a firm's market share and return on assets[402]. In a meta-analysis about studies published from 1997 to 2010, Luo et al. found a significant, positive association with organizational performance and government ties[403]. Government officials can provide support and incentives for starting new businesses, as well as develop a friendly business environment, such as through the ease of renting facilities[404]. They can also support to procure materials at lower costs, provide priority access to infrastructure or promote products in state-controlled distribution channels[405]. As government officials approve applications, for instance for advertisements, and grant rights such as import licenses, Guanxi with them can furthermore help to facilitate approvals[406]. As well-established institutional law is considered still absent, and the government's mode of action frequently opaque, a good relationship with appropriate authorities is crucial to get certain permissions granted and to avoid long delays[407]. This is particularly important for foreign direct investment ventures that enter China[408]. Guanxi with the government can also help to improve company reputation for critical stakeholders[409]. Reputation is the "long-term combination of outsiders' assessments about what the organization is, how well it meets its commitments and conforms to stakeholders' expectations, and how effectively its overall performance fits with its sociopolitical environment[410]." Based on the Chinese understanding of government leaders as almost "divine figures" with a mandate to rule, official recognition from officials, for instance

---

[397] Luo et al. (2014), pp. 105-107; Geng et al. (2017), p. 3.

[398] The term "government" refers to all government bodies in China at different levels. Central government (the State Council and responsible ministries or bureau), local provincial or city government and its responsible departments, industry or trade associations (many of them are either former government agencies or have close links with ministries), Communist Party organizations and trade unions affiliated to the Party and other government departments that may indirectly affect business operation (e.g. public security bureau), cf. Fan (2007), pp. 504-505.

[399] Davies et al. (1995), p. 211.

[400] Gu et al. (2008), p. 17.

[401] Chen, Ming-Huei et al. (2015), p. 901.

[402] Peng/Luo (2000), pp. 494-495.

[403] Luo et al. (2012), pp. 158-161.

[404] Chen, Ming-Huei et al. (2015), p. 902.

[405] Peng/Luo (2000), p. 495.

[406] Davies et al. (1995), p. 211.

[407] Luo et al. (2012), p. 147; Tung/Worm (2001), pp. 521-522.

[408] Jiang (2006), p. 891.

[409] Davies et al. (1995), p. 211; Lee et al. (2018), pp. 357-358.

[410] Brown/Logsdon (1999), pp. 323-334.

in the form of pictures where business figures shake hands with high-status officials, is a very effective way to convey a firm's status and prestige[411].

4.5. Costs

While Guanxi can have certain benefits when applied with other companies or government officials, it also comes at costs, which will be introduced in this section.

Guanxi with other companies

Developing and maintaining Guanxi is time intensive so that only a few Guanxi relationships can be maintained[412]. Given Guanxi's personal nature, relationships are developed person-to-person, involving human interaction such as joint meals, golf, or spending time at Karaoke bars[413]. In a survey by Yi and Ellis, most respondents highlighted time consumption as a key disadvantage of Guanxi[414]. Establishing and retaining Guanxi is also expensive, and this investment might not be cost-efficient[415]. Guanxi ties with other firms can be so costly that they only lead to increased profitability under certain circumstances[416]. Guanxi development also involves both favors and obligation, where an obligation is created whenever a gift, banquet, or favor is accepted[417]. Rejecting the repayment of favors is regarded as immoral so that they should be repaid with at least equal value[418]. The receiver of a favor is not expected to return an owed favor to a benefactor immediately, but instead "store the favor for as long as it takes," and return it once the other party needs it[419]. This can become especially challenging when the expected favor in a business context would lead to the violation of duties[420].

Guanxi with government officials

Like Guanxi with other companies, developing and maintaining Guanxi with government officials is time-intensive and requires personal effort[421]. In addition, cultivating Guanxi with government officials is also expensive, especially as a payment of "management fees" is often expected[422]. Warren et al. showed throughout studies with Chinese business people that the high costs associated with Guanxi building activities are a vital issue just as the expectation to bribe government officials with money, prostitutes, or by conducting smuggling activities[423]. These costs are often so extensive that overall political ties negatively affect a firm's profitability[424]. Furthermore, indebtedness and obligation out of the reciprocity of favors can become a severe liability for an organization[425]. Guanxi with government officials leads to demand requests and interventions such as improving the local employment rate or the recruitment of unqualified employees who are relatives of some government leaders in key

---

[411] Fan (2007), p. 503.
[412] Fan (2002b), p. 547; Ewing et al. (2000), p. 84.
[413] Styles/Ambler (2003), p. 638.
[414] Yi/Ellis (2000), pp. 28-29.
[415] Fock/Woo (1998), p. 38; Yi/Ellis (2000), pp. 28-29; Adler/Kwon (2009), p. 106.
[416] Li/Sheng (2011), pp. 565-567; Li et al. (2008), pp. 392-395; Peng/Luo (2000), pp. 494-497.
[417] Cheng et al. (2012), p. 7; Mei-Hui Yang (1994), p. 142.
[418] Hwang (1987), p. 957; Zhang/Zhang (2006), p. 381.
[419] Yen et al. (2011), p. 99.
[420] Dunfee/Warren (2001), pp. 200-201.
[421] Fock/Woo (1998), p. 38; Ewing et al. (2000), p. 84; Styles/Ambler (2003), p. 638.
[422] Park/Luo (2001), p. 462.
[423] Warren et al. (2004), pp. 355-364.
[424] Li et al. (2009), pp. 347-348.
[425] Vanhonacker (2004), p. 49.

positions[426]. For foreign companies investing in China, this can also mean that they are requested to form joint ventures with government-related partners[427]. Although the CCP published and enforced ethical rules to control certain Guanxi practices and various anti-corruption campaigns were launched, Guanxi with officials might today still be perceived as corruption[428]. Giving gifts to government officials is often done with the intent to receive an illegitimate benefit later. The impression of corruption is also aggravated as traditionally luxury watches, lavish banquets, prostitutes, large sums of money, or houses were provided in exchange for favorable treatment within an official's sphere of authority[429].

4.6. Guanxi-corruption relation

The relationship between Guanxi and corruption has been debated among scholars for a long time[430]. In this section, corruption will be defined, and its effects explained, the history of corruption in China outlined, and the intertwinement of Guanxi and corruption discussed.

Corruption definition and effects

Defining corruption is a challenge as meanings can be depending on the specific social and political context in which the word is used[431]. Therefore, definitions are often too general to be operable or too narrow, making them only applicable to specific, well-defined cases[432]. The most popular used one is "the misuse of public office for private gain[433]." Another important definition approach adds a social component and therefore states that corruption is "behavior of public officials which deviates from accepted norms in order to serve private ends[434]." Hence, corruption must include some illicit behavior, deviating from norms accepted in the specific cultural and reference groups, such as politicians or civil servants[435]. Based on these and further definition approaches used in the literature, Philp concludes that a case of corruption exists when "a public official (A), acting for personal gain, violates the norms of public office and harms the interests of the public (B) to benefit a third party (C) who rewards A for access to goods or services which C would not otherwise obtain[436]."

Evaluating the prevalence of corruption is challenging because it often occurs in secret, and violence or intimidation might be used to "see off investigators and keep others quiet[437]." Therefore, typically surveys about the perception of corruption prevalence are employed[438]. Here, issues arise with the choice of respondents as well as an incorporation of personal prejudices[439]. Studies show that corruption is perceived to be more widespread in federal states, less developed economies, and countries with low imports, while countries with Protestant

---

[426] Li et al. (2009), pp. 343-344; Warren et al. (2004), p. 366.

[427] Abramson/Ai (1999), p. 13.

[428] Harding (2013), p. 127; Fock/Woo (1998), p. 38.

[429] Harding (2013), p. 127; Fock/Woo (1998), p. 38.

[430] Li (2011), p. 3; Fock/Woo (1998), p. 38.

[431] Sampford (2006), p. 57.

[432] Waite/Allen (2003), p. 282.

[433] Brown (2006), p. 65; Treisman (2000), p. 399.

[434] Huntington (1968), p. 59.

[435] Philp (2006), p. 47.

[436] Philp (2002), p. 42; Philp (2006), p. 45.

[437] Kim/Lachapelle (2000), pp. 3-7.

[438] Commonly used surveys are Transparency International's "Corruption Perceptions Index" or the World Bank's "Control of Corruption index," cf. Transparency International (2020), pp. 1-29; The World Bank Group (2018b).

[439] Tansey (2007), pp. 768-770; Philp (2006), pp. 49-51. Donchev and Ujhelyi also presented evidence of systematic biases in such studies, including a diminishing sensitivity to experiences, that is low levels of corruption experience having a disproportionate impact on perceptions, influences of absolute instead of relative levels of corruption and effects through attitudes depending on country characteristics, cf. Donchev/Ujhelyi (2014), pp. 309-331.

traditions, a history of British rule, or extended exposure to democracy are seen as less corrupt[440]. Heywood also argues that as since the 1990s, Western civilization is facing a challenge from alternatives of Islamic and Asian states, growing corruption and decadence becomes symptomatic of Western civilization in decline[441]. Alatas believes that corruption is a phenomenon found everywhere, although its nature, causes, and functions might differ from country to country[442]. Huntington also argues that as corruption is one measure of the absence of effective institutions, it is prevalent in any society that undergoes intense modernization phases[443].

Corruption leads to undermining a society's "fairness, stability, and efficiency" and its ability to deliver sustainable development to its members[444]. For instance, contracts might not be awarded to the most efficient producer, and new producers' entry might be blocked[445]. Costs of transactions might be increased, which reduces industrial policies' effectiveness, and rates of foreign direct investments can be harmfully impacted[446]. Higher levels of corruption are also associated with a reduction in the productivity of public investments. Uneconomic "white elephant" projects are undertaken, the quality of infrastructure decreases and the reduced government income harms the capability to finance productive growth[447]. Rising corruption also comes along with an increase in income inequality and poverty[448]. Besides, the cost of capital for companies in countries with high corruption is raised, as investors must reassess possible returns of investments, incorporating the risk of corruption induced changes in prices and production factors[449]. Furthermore, entrepreneurs might be incentivized to allocate their talent to become rent-seekers that add little economic value[450]. While these definition approaches and impacts of corruption focus on the public sector, corruption and related behavior also exist in a business setting[451]. The Association of Certified Fraud Examiners (ACFE) lists corruption as one element in its occupational fraud and abuse framework. Here, corruption occurs when "fraudsters wrongfully use their influence in a business transaction in order to procure some benefit for themselves or another person, contrary to their duty to their employer or the rights of another," and corruption schemes are broken down into bribery, illegal gratuities, economic extortion and conflicts of interest[452].

Corruption in China

Corruption is a challenge in China for centuries. Already in the era of emperors and Confucius, officials' values were a concern. Later in the early 20th century, China's economy was weakened by government and military corruption, and the fall of the Kuomintang regime in 1949 is partly attributed to rampant corruption during that time[453]. In Mao Zedong's regime, strict rules had to be introduced to restrain corrupt behavior and bureaucratic misconduct[454]. However, since

---

[440] Treisman (2000), p. 399.
[441] Heywood (1997), pp. 430-435.
[442] Alatas (2015), ix.
[443] Huntington (1968), pp. 59-71.
[444] Shacklock et al. (2006), p. 1.
[445] Seligson (2002), pp. 409-410; Rose-Ackerman (1997), pp. 42-46.
[446] Ades/Di Tella (1997), pp. 1038-1041; Wei (1997), pp. 1-27.
[447] Shacklock et al. (2006), p. 1; Tanzi/Davoodi (1998), pp. 55-58.
[448] Gupta et al. (2002), pp. 30-42.
[449] Murphy et al. (1991), pp. 503-530; Gray/Kaufmann (1998), pp. 7-8.
[450] Murphy et al. (1991), pp. 503-530; Jain (2001), p. 97.
[451] Tanzi (1998), pp. 564-565.
[452] Wells (2014), pp. 44-272.
[453] Hao/Johnston (1995), pp. 80-94.
[454] Dai (2013), pp. 61-76.

the economic liberalization in 1978, intense transitions of political and economic institutions have created enormous opportunities for corruption[455]. New connections between wealth and power led to rampant corruption, making it a threat to social and political stability in the 1990s[456]. Corruption increased in quantity and intensity, visible in a higher share of major economic offenses and a higher involvement of senior CCP cadres and government officials in illegal activities[457]. The Chinese government and the CCP tried to address corruption with the launch of various anti-corruption campaigns. Since 1982, economic crimes, violations against party discipline, engagement in corrupt activities, graft, bribery, and other relevant crimes were prosecuted[458]. The latest major anti-corruption campaign was launched by the Chinese President Xi Jinping and his administration, only twenty days after taking office in 2012, in the form of an "Eight-point Policy," demanding government officials to stop accepting extravagant perks[459]. Xi Jinping warned that corruption could lead to a downfall of the state, and therefore the campaign is designed to deter bureaucrats from even considering any corrupt behavior[460]. Although these efforts showed some success in reducing mostly low-level corruption, they are considered overall inadequate[461]. In Transparency International's 2019 Corruption Perceptions Index, China still ranks on 80th place out of 180 considered countries[462]. Dai suggests that this is for various reasons[463]. In the past decades, the elite has been integrated into a privileged network of shared interests. So, although anti-corruption measures had to be implemented to "pacify the anger of common people" and maintain the regime's moral image, they were never aimed at sincerely harming the interests of these stakeholders. Also, institutions and political structures that are barriers to anti-corruption efforts are not permitted to change genuinely. Finally, rules, laws, and regulations still contain too many loopholes and weaknesses, which allows for manipulation[464].

Intertwinement of Guanxi and corruption

For corruption to flourish, a human interface is needed to overcome legal, moral, and cognitive barriers that would otherwise block such behavior[465]. Guanxi networks with their communication, exchange, and normative functions support breaking down practical and psychological hurdles and therefore facilitating corruption[466]. This is also acknowledged by other scholars that argue that Guanxi ties may not remove the corruption problem in networks or markets or are even a necessary and integral part of the corruption in China[467]. Guanxi is seen as an additional factor weakening China's corporate governance system, thus harming its international economic competitiveness[468].

A contradicting view is taken by scholars that argue that Guanxi is different from corruption. Ambler explains that Guanxi is "no more equivalent to corruption than social drinking is to

---

[455] Zhan (2012), pp. 99-105.
[456] Hao/Johnston (1995), pp. 80-94.
[457] Wederman (2004), pp. 920-921.
[458] He (2000), pp. 267-268.
[459] Lin et al. (2016), p. 2.
[460] Lin et al. (2016), p. 2; Leung (2015), p. 32.
[461] Wederman (2004), pp. 920-921; He (2000), p. 269; Guo (2008), pp. 349-364; Gong (2002), pp. 85-103.
[462] Hao/Johnston (1995), pp. 80-94; Wederman (2004), pp. 920-921; Transparency International (2020), pp. 2-3.
[463] Dai (2013), pp. 61-76.
[464] Dai (2013), pp. 61-76.
[465] Li (2011), p. 20.
[466] Zhan (2012), pp. 99-105.
[467] Gao et al. (2012), p. 464; Li (2011), p. 20.
[468] Braendle et al. (2005), pp. 399-402.

drunkenness[469]." Guanxi can then be seen as a civil culture that complements the overall legal system[470]. It differs from corruption in various aspects[471]. It is an ingredient of China's social norm, while corruption is, per one definition, a deviation from the social norm. Guanxi is typically legal, while corruption is clearly illegal. It does not involve any legal risks if it fails, while corruption is linked to legal risks. Guanxi is built on interpersonal favor exchanges, corruption mostly on monetary exchange. Also, Guanxi involves implicit, social reciprocity, while corruption relies on explicit, transactional reciprocity. Guanxi is set for long-term orientation; corruption is typically built on short-term transactions. Corruption is based on the exchange between money and power, while Guanxi is founded on trust. Finally, Guanxi is transferable, whereas corruption is not[472].

An integrative approach of these two viewpoints is taken by other scholars that distinguish between Guanxi to "establish good business relations" and Guanxi practice, which is "using social relations to take care of procedures" and consider only the latter corrupt and taboo[473].

4.7. Cultural uniqueness

There is disagreement among researchers to which extent Guanxi exists in all cultures, in certain cultures, or is something unique to China. These three perspectives will be discussed in this section.

Existing in all cultures

Some scholars argue that Guanxi's key attributes, such as trust, mutual obligation, and reciprocity, are universally practiced as informal facilitation of formal processes in business, politics, and society[474]. For instance, the notion of trust was also theorized in Western literature as a precursor for successful relationships[475]. Zand showed that trust is an essential factor of organizational performance as it supports information sharing, the acceptance of mutual influence, and inspires self-control[476]. Morgan and Hunt demonstrated that commitment and trust lead to cooperative behavior and produce outcomes that promote efficiency, productivity, and effectiveness[477]. Also, the norm of reciprocity is considered a base to establish social relationships effectively among any human beings[478]. To requite a benefit, or to be grateful to someone who bestows it, is at least to some degree universally regarded as a duty[479]. The norm of reciprocity can be seen as omnipresent, notably the demands that "people should help those who have helped them" and "people should not injure those who have helped them" are found in all value systems[480]. Lisha et al. also support the Guanxi's universality in different cultures, arguing that reciprocity in the form of gift-giving exists in most social relationships across all societies[481]. They also hypothesize that the concept of face has overlapping elements with

---

[469] Ambler (1994), p. 75.
[470] Lo/Otis (2003), p. 143; Potter (2002), pp. 194-195.
[471] Luo (2007), pp. 215-218; Li (2011), p. 3; Chen/Chen (2004), p. 309.
[472] Luo (2007), pp. 215-218; Li (2011), p. 3; Chen/Chen (2004), p. 309.
[473] Guthrie (1998), p. 255; Gold et al. (2002), p. 6.
[474] Qi (2013), p. 321.
[475] Mavondo/Rodrigo (2001), p. 114.
[476] Zand (1972), pp. 235-238.
[477] Morgan/Hunt (1994), pp. 22-24.
[478] Hwang (1987), p. 956.
[479] Westermarck (1908), pp. 153-155.
[480] Gouldner (1960), p. 171.
[481] Lisha et al. (2017), pp. 1128-1129.

dignity and that ganqing is similar to emotional ties in Western social networking[482] . Some scholars argue that Guanxi is, therefore, merely the Chinese form of relationship marketing[483].

Existing in some cultures

A different perspective is propagated by some scholars that believe that Guanxi exists in some cultures that are like the Chinese.

Guanxi is considered ubiquitous in all Confucian societies and equal to Kankei (Japanese "関係," the subconscious notion of granting access through relationship), in Japan or Kwankye (Korean "관계," relations) in Korea[484]. Confucianism is a philosophy of "human nature," where proper human relationships are seen as society's base. The cardinal principle is "ren" (Chinese "仁," translated as "humanity"), which refers to "warm human feelings between people[485]." Confucian philosophy highlights that individuals are not "isolated entities" but part of a larger interdependent relationship system. Consequently, countries that were affected by this philosophy focus on building and managing effective relationships[486]. Guanxi is therefore perceived as the "East Asian form of business connection," consisting of the maintenance of relationships with the proper organizations and individuals within them[487].

Other scholars understand Guanxi, as it is based on family and kinship ties, as social capital occurring in all collectivist cultures[488]. In Hofstede's framework for understanding national differences, collectivism, as opposed to individualism, describes the relationship between the collectivity and the individual of a society[489]. In collectivist cultures, societal norms include a "we consciousness," the belief that in a society, people are born into extended families or clans, which protect them in exchange for loyalty, that value standards are different for people in a group and outside of it, and that expertise, order, duty, and security are provided by an organization[490]. Social capital in these cultures is a resource arising from strong community ties, especially where legal systems and enforcement mechanisms are weak[491]. Examples include Ubuntu (Bantu, translated as humaneness), a "pervasive spirit of caring and community, harmony and hospitality, respect and responsiveness that individuals and groups display for one another," in South Africa or in Russia Blat (Russian "блат"), a word used during socialism when referring to the exchange of favors to get access in conditions of shortages at public expense[492].

Existing only in China

Another school of thought believes that although some commonalities of Guanxi can be found in other cultures, at least nuanced differences to other cultural concepts exist[493]. A comparison between the role of Guanxi in China and relationships in Australia in direct selling found that

---

[482] Lisha et al. (2017), pp. 1128-1129.

[483] Ballantyne (1994), pp. 2-6; Davies et al. (1995), pp. 208-213. For a review about the concept of relationship marketing including an introduction of key differences between relationship marketing and transactional marketing, cf. Hennig-Thurau/Hansen (2010), pp. 3-12.

[484] Hitt et al. (2002), p. 358; Yang/Wang (2011), p. 492; Usunier/Lee (2005), p. 464; Yeung/Tung (1996), p. 63.

[485] Li (2007), p. 311; Yum (1988), p. 377.

[486] Hitt et al. (2002), p. 358.

[487] Usunier/Lee (2005), p. 464.

[488] Gu et al. (2008), p. 15.

[489] Hofstede (2001), p. 209.

[490] Parsons/Shils (2017), p. 82; Hofstede (2001), p. 227.

[491] Berger/Herstein (2012), pp. 29-30.

[492] Mangaliso (2001), p. 24; Michailova/Worm (2003), p. 510; Berger et al. (2015), p. 168; Gu et al. (2008), p. 15.

[493] Berger/Herstein (2012), pp. 29-30; Shi et al. (2011), p. 456; Hwang (1987), p. 956; Lisha et al. (2017), pp. 1128-1132.

although in both countries, relationship elements affected the selling effectiveness, the impact was much more substantial in China[494]. A key difference was that relationship marketing in China was holistically rooted in Guanxi, while in Australia, it was separated instead of integrated into dimensions of friendship, trust, and reciprocity[495]. Wang also compared Western relational marketing with Guanxi, concluding that Guanxi is more personal, includes an emotional attachment, and is highly network-specific[496]. Gu et al. contrasted relational exchange and Guanxi networks as a construct as well as a governance structure. Guanxi's differences include its usage as a corporate resource, its high application frequency to many facets of life, its antecedents based on bounded solidarity, and the maintenance process through accumulated obligations and reciprocity[497]. When studying managerial principles in Japanese, Chinese, and Korean business organizations, Alston concluded that each society has its separate, distinctive philosophy that guides managers there. In China, personal relations are in focus; in Japan, group harmony and social cohesion are most prominent, and in Korea, a concern for harmony based on the respect of hierarchical relationships is dominating[498].

---

[494] Merrilees/Miller (1999), p. 267.
[495] Merrilees/Miller (1999), p. 272.
[496] Wang (2007), pp. 81-86.
[497] Gu et al. (2008), p. 14.
[498] Alston (1989), p. 26.

5. Expert interviews to identify the impact of Guanxi on the LSP selection in China

Chapter 3 introduced logistics outsourcing and its financial as well as strategic and operational benefits. A critical success factor in unlocking these benefits are BSRs, which are managed by boundary spanners. Chapter 4 then discussed the phenomenon of Guanxi. Guanxi ties can lead to certain benefits and costs for a company in China. Studies about Guanxi and Guanxi practices in a business setting typically take a macro view. Few take a micro-view, and so far, none addressed the impact of Guanxi on the selection of LSPs. As illustrated in Figure 6, based on the previously introduced concepts, the exploratory study in this chapter will close the research gap by answering RQi1, how does Guanxi impact the LSP selection in China.

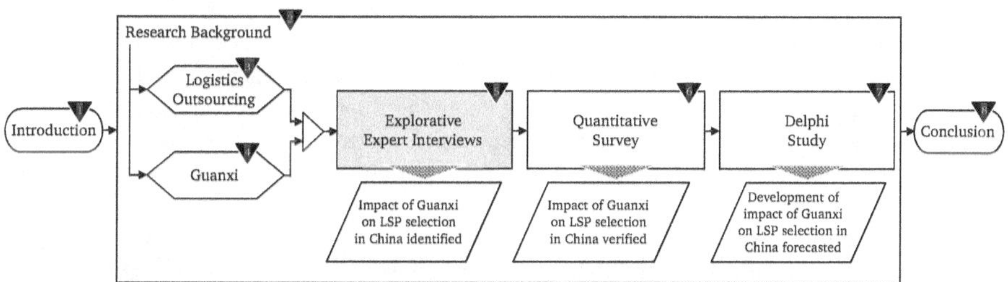

Figure 6: Chapter "Explorative Expert Interviews" within overall research process
Source: Own figure

For this, section 5.1 will describe the current state of research about Guanxi of boundary spanners in the selection of LSPs. To answer RQi1, the methodology of qualitative expert interviews was used. Section 5.2.1 will explain the background of this technique and evaluate it. In section 5.2.2, the study procedure is then laid out, including the points that must be considered when conducting expert interviews in an intercultural setting about a sensitive topic and the interview guide's design. Afterwards, the selection process for the participants and the structure of the respondents are described throughout section 5.2.3. The analysis process for the around 120 pages of interview transcripts and notes that could be gathered is then briefly introduced in section 5.2.4. Section 5.3 then summarizes the results in section 5.3.1 and discusses them. The identified eight support possibilities and ten contingency factors are explained in detail and related to the current research state throughout section 5.3.2 and 5.3.3. Furthermore, limitations and possible endeavors for further research are pointed out in section 5.3.4.

5.1. Guanxi of boundary spanners in the selection of LSPs

Logistics outsourcing, contracting out of logistics activities previously performed in-house, is increasingly common to realize financial, strategic, and operational benefits[499]. By employing LSPs, companies expect lower logistics costs due to increased efficiency and better-utilized assets, fewer capital expenditures, increased service levels, flexibility, higher customer

---

satisfaction, and the possibility of focusing on their core competences[500]. Selecting LSPs is a multi-dimensional decision of high importance for companies where various tangible and intangible criteria, such as cost, quality, reliability, flexibility, compatibility, services, financial measures, sustainability measures, or delivery, must be considered[501].

China's logistics market is still developing and faces various challenges such as low efficiency and high logistics cost, congestion, lack of integrated intermodal transport networks, local protectionism, inadequate IT infrastructure, and underdeveloped warehousing services[502]. However, these inefficiencies are anticipated to change due to government initiatives intended to upgrade China's logistics network and improve its capacity and efficiency[503]. Along with this development, just like in the rest of the world, logistics outsourcing is expected to become more common[504].

Guanxi, often attempted to be translated as (special) relationship, connections, or particularistic ties, is a unique phenomenon in China that impacts the whole society[505]. In a business setting, Guanxi can help to gain valuable and relevant information, enhance business, decrease transaction costs, and prevent opportunistic behavior[506]. However, there are also disadvantages and risks connected with it. Establishing and maintaining Guanxi is time-intensive so that only a few Guanxi relationships can be maintained, and expensive[507]. As Guanxi development involves favors and an obligation to repay favors, it can become a critical liability, especially when duty violations are demanded[508].

Boundary spanners are people who operate at an organization's boundary and manage the interface between organizations and their environments as linking pins or gatekeepers[509]. They perform an information processing and an external representation or gatekeeping function[510]. In their information processing function, they search, filter, summarize information, and facilitate information dissemination[511]. In their external representation or gatekeeping function, they acquire and dispose resources, political legitimacy and hegemony, social legitimacy, and project the organizational image[512]. Challenging for the involvement of boundary spanners is that they operate under their employer's name, which in return, must trust these employees to fulfill their role conscientiously. A high degree of freedom for boundary spanners can lead to improved outcomes but allows for opportunistic behavior that can harm a company[513].

Guanxi can influence supplier search mechanisms when companies use the Guanxi of their employees in a business context[514]. Especially for new entrants to the Chinese market, the use of Guanxi might be beneficial to find suppliers, obtain marketing channels, and improve sales

---

[500] Wilding/Juriado (2004), p. 630; Sink/Langley Jr (1997), p. 182; van Damme/Amstel (1996), p. 88; Bask (2001), pp. 484-486; Selviaridis et al. (2008), p. 385.

[501] Alkhatib et al. (2015), p. 133; Alkhatib et al. (2015), p. 105; Aguezzoul (2014), pp. 69-78.

[502] Zhang/Figliozzi (2010), pp. 179-190; Hong et al. (2004), p. 20.

[503] Economist Intelligence Unit (2014); Jiang (2018), pp. 1-24.

[504] Chu/Wang (2012), pp. 78-79.

[505] Yeung/Tung (1996), p. 55; Alston (1989), p. 28; Ai (2006), p. 105; Jacobs (1979), p. 238; Gold et al. (2002), p. 3.

[506] Shin et al. (2007), pp. 171-172; Chen, Ming-Huei et al. (2015), p. 903; Wiegel/Bamford (2015), p. 315; Peng/Luo (2000), p. 494; Luo et al. (2012), pp. 158-161; Standifird/Marshall (2000), p. 30; Davies et al. (1995), pp. 208-213; Lee/Humphreys (2007), pp. 458-464; Wang (2007), p. 85.

[507] Fan (2002b), p. 547; Fock/Woo (1998), p. 38; Yi/Ellis (2000), pp. 28-29; Ewing et al. (2000), p. 84; Styles/Ambler (2003), p. 638.

[508] Cheng et al. (2012), p. 7; Hwang (1987), p. 957; Zhang/Zhang (2006), p. 381; Dunfee/Warren (2001), pp. 200-201.

[509] Organ (1971), p. 73; Leifer/Delbecq (1978), pp. 40-41; Katz/Kahn (1978), 17-743; Utterback (1971), p. 84.

[510] Adams (1976), pp. 1175-1199.

[511] Aldrich/Herker (1977), pp. 218-221.

[512] Aldrich/Herker (1977), pp. 218-221.

[513] Perrone et al. (2003), pp. 434-435; Luo (2006a), pp. 141-144.

[514] Millington et al. (2006), p. 527.

as well as business performance[515]. In a logistics context, the boundary spanning properties of Guanxi are believed to have a positive impact on the quality of a buyer-supplier relationship[516]. Relationships can have an impact on operational performance and sales opportunities. On an operational level, Guanxi has a positive impact on logistics service value and the service quality of physical distribution[517]. Regarding the cooperation with third-party logistics providers, Chen et al. analyzed an increase in business services flexibility and hypothesized a positive effect due to Guanxi, however, could not demonstrate this link directly[518]. Researchers mainly took a macro view on Guanxi and Guanxi practices in a business setting, while only few took a micro-view. Also, while logistics outsourcing is an increasingly important field in China, Guanxi's impact on the selection of LSPs was so far not addressed. Studies about the Guanxi influence in supply chain management made a variety of noteworthy contributions. For instance, due to the favor exchanging nature of Guanxi, it could be shown that it can support risk management and supplier development activities among supply chain partners[519]. It was also revealed that Guanxi fosters strategic purchasing and can facilitate outsourcing by creating an alternative governance system that reduces transaction costs[520]. Finally, the strong ties between a buying and supplying firm in a Guanxi relationship can support the introduction of GSCM initiatives[521]. Asides, studies about BSR showed that during the supplier selection, buyers' choices are impacted by bonded relationships with their existing suppliers[522]. Mummalaneni et al. discovered that the relationship with a supplier, as well as characteristics of the salesperson, are evaluation criteria during the supplier selection, especially in China[523]. Also, Kamann and Bakker found that the selection process of suppliers and the contents of supplier relations are subject to certain subjective perceptions[524]. Huang et al. could demonstrate that interpersonal trust is formed among boundary spanners from different organizations during a supplier selection process[525]. Kannan and Choon Tan analyzed the effect of the supplier selection process on relationship performance and found that it positively affects a buyer's performance[526]. Similarly, Carey et al. stretched the importance of social and relational factors when forming relationships between industrial buying firms and their key suppliers, as it translates into long-term performance benefits[527]. Social bonds can hold associates from buying and selling firms personally together, which can act as exit barriers that prevent ending a relationship despite an overall low satisfaction level[528]. Based on a comprehensive literature review and a focus group study in the US, Carter showed that various ethical issues, such as writing specifications that favor a particular supplier or allowing only certain suppliers to bid, can be involved in the relationship of purchasing managers and their suppliers[529].

This is particularly relevant for the logistics outsourcing industry, which is characterized by partnership-type network relationships and high levels of customization, which are associated

---

[515] Luo/Chen (1997), pp. 14-15.
[516] Chen et al. (2010), pp. 289-290.
[517] Chao/Anantana (2014), pp. 92-95; Towers/Xu (2016), pp. 134-135.
[518] Chen, Haozhe et al. (2015), p. 263.
[519] Lee/Humphreys (2007), pp. 455-462; Lee/Humphreys (2007), pp. 462-463.
[520] Lee/Humphreys (2007), pp. 462-463; Davies et al. (1995), pp. 208-213.
[521] Simpson et al. (2007), pp. 42-44; Cheng et al. (2012), p. 11.
[522] Kamp (2005), pp. 660-665.
[523] Mummalaneni et al. (1996), pp. 120-122.
[524] Kamann/Bakker (2004), pp. 62-63.
[525] Huang et al. (2008), pp. 55-71; This is also related to the phenomenon of deception in BSR negotiations, cf. Rottenburger (2019), pp. 3-5.
[526] Kannan/Choon Tan (2006), pp. 757-763.
[527] Carey et al. (2011), pp. 285-286.
[528] Gounaris (2005), pp. 129-136; Jonsson/Zineldin (2003), pp. 229-237.
[529] Carter, Craig R. (2000), pp. 191-206.

with high levels of interdependence and complexity in the form of uncertainty, difficulty, and variability[530]. Therefore, the value of relationship marketing is very high in the logistics outsourcing industry, where relational performance by LSPs is often considered as the single most important factor in creating customer satisfaction[531]. Chu and Wang analyzed relationship quality in the context of logistics outsourcing and found that higher relationship quality, measured in benevolence trust, capability trust, commitment, and satisfaction, resulted in increased financial performance[532].

In Asia, notably China, the logistics industry is notorious for its closeness to relationship-based practices. CEOs of the largest LSPs operating in Asia-Pacific indicated that corruption, a practice in China often associated with Guanxi, is among the Top 3 problems faced by their companies in the region[533]. Complicated customs procedures and excessive customs clearing times for foreign companies, attributed to non-procedural practices at customs and poorly defined scope of administrative and discretionary powers vested in custom officials, are also considered as related to Guanxi[534].

Considering the impact of Guanxi on supply chain management, the effect of personal relationships on the selection of and cooperation with suppliers, the logistics outsourcing industry's particularities, and the characteristics of business relations in China, it is likely that Guanxi also impacts the selection of LSPs[535]. This is also reflected in the author's personal experiences while working in the Central Logistics department of a Tier 1 automotive supplier in China. During LSP tender processes, some LSPs had information beyond what was shared through the official communication channels. A few possessed detailed knowledge about individual plant logistics managers' unofficial wishes regarding operations and acceptable costs. Some key account managers of LSPs also knew about meetings previously held with their competitors.

This study aims to identify possible impacts of Guanxi on the selection of Logistics Service Providers in China. Therefore, research question RQi1 states:

> *RQi1: How does Guanxi impact the LSP selection in China?*

Such knowledge can be described as process knowledge that can best be discovered through semi-structured expert interviews[536].

5.2. Method: Qualitative Expert Interviews

In this section, the methodology "Qualitative Expert Interviews" will be introduced and evaluated. Then, the study procedure laid out, and the participant selection and their profile presented. Finally, the analysis approach will be briefly described.

5.2.1. Background and evaluation

Expert interviews are a very popular, widely used research method. Bogner et al. explain that this is mainly because talking to experts, for instance in the exploratory phase of a research

[530] Vickery et al. (2004), 1107-1118.
[531] Knemeyer/Murphy (2004), pp. 36-47.
[532] Chu/Wang (2012), pp. 78-86.
[533] Lieb (2008), pp. 506-507.
[534] Chan (2001), pp. 522-527.
[535] Chen, Haozhe et al. (2015), p. 263.
[536] Bogner et al. (2014), p. 18.

project, is an efficient data gathering method[537]. Through experts' practical knowledge, the interviewer can quickly get a thematic orientation and understand potential problems, challenges, and facts [538] . Therefore, the time-consuming data collection process can be shortened, especially when experts can represent a wider group of people and act as "crystallization points" for practical insider knowledge[539]. They also have a high potential to obtain usable information that would otherwise be difficult to access, such as taboo subjects[540]. Expert interviews regularly lead to high-quality results, as experts often feel motivated to participate in an interview when the interviewer and interviewee share a common scientific background or relevance system. Besides, experts tend to like to help to "make a difference" and therefore support scientific research about current topics. Furthermore, experts are frequently professionally curious about relevant research in their field of interest and desire to share their thoughts and ideas with an external specialist[541]. However, expert interviews are sometimes linked to longstanding biases about qualitative research[542]. Critiques consider them "too subjective" and "susceptible to human error and bias in data collection and interpretation[543]." Due to non-probability selection of experts and an unclear definition of what constitutes an expert, some also describe them as lacking representativeness. In addition, while it is acknowledged that methodological concepts and suggestions can be helpful, not only precisely one correct way to conduct expert interviews exists. Therefore, the research design must be context-based, which can be criticized as a nonsystematic design[544]. Nevertheless, the method expert interviews should not be slighted as it can save time and money and generate findings that are particularly suitable as a base for further analyses and hypotheses[545].

The Expert interview is a guided variant of the research interview in which the interviewees are experts on a topic, and their specialist knowledge - typically structural expertise or practical knowledge - should be accessed and collected[546]. A research interview is a targeted, systematic, and rule-based generation and recording of verbal statements of either one interviewee (individual survey) or several interviewees (group survey) on selected aspects of their knowledge, experience, and behavior, in oral form[547]. Interviews can be conducted face-to-face, by telephone, or online. An interviewer provides the verbal questions that an interview is based on to interviewees in one interaction. Answers are documented and systematically analyzed. The research interview's central elements are the interviewee(s), the interviewer, the interview situation, and the interview questions[548].

Based on their purpose, expert interviews can be distinguished in exploratory, systematizing, and theory-generating. *Exploratory* expert interviews are used as initial orientation in a substantively new or poorly defined field to help a researcher to develop a clearer understanding of an issue or as the first step to develop questions for an interview guide[549]. Here, expert interviews can help to structure an area under investigation and generating hypotheses. Experts can be part of the study target group or a complementary source with

[537] Bogner et al. (2009), pp. 1-3.
[538] Bogner et al. (2014), pp. 64-67.
[539] Bogner et al. (2009), pp. 1-3.
[540] Bogner et al. (2009), p. 2; Döring et al. (2015), pp. 151-153.
[541] Bogner et al. (2009), pp. 1-3.
[542] Cooper/Schindler (2014), p. 94.
[543] Cooper/Schindler (2014), pp. 144-145.
[544] Kassner/Wassermann (2002), pp. 109-110.
[545] Cooper/Schindler (2014), p. 129; Bogner et al. (2014), pp. 64-67.
[546] Döring et al. (2015), p. 376.
[547] Döring et al. (2015), p. 356.
[548] Döring et al. (2015), p. 356.
[549] Bogner/Menz (2009), pp. 46-48.

information about the actual study group. The exploratory interview's focus is to question the subject under investigation, acquiring as much information as possible or standardizing data[550]. *Systematizing* expert interviews are oriented towards gaining access to exclusive knowledge possessed by an expert. The target is to obtain such knowledge that has been derived from practice and is rather spontaneously accessible and communicated. This type of expert interview focuses on obtaining systematic and complete information. As their specialized knowledge and information pieces are not available to researchers, the experts enlighten them like guides[551]. In *theory-generating* expert interviews, an expert does not merely serve as an accelerator for the research process, by which a researcher can obtain useful information about an issue under investigation. Instead, the goal of the theory-generating interviews is to communicatively open up and analytically reconstruct the subjective dimensions of expert knowledge. Based on subjective behavior and implicit decision-making maxims of experts as a starting point, a theory should be formulated[552].

5.2.2. Procedure

As expert interviews are not sufficiently standardized to meet the requirements of intersubjective verifiability, the process of data collection, analysis, and interpretation must be transparently documented[553]. This section will describe points that must be considered when conducting expert interviews and the design approach for the interview guide.

Interview conduction

Interviewers should possess high *communication and social skills* as well as the ability to self-reflect. This is necessary to create a pleasant conversation atmosphere - that can also be maintained when complications occur - with a wide variety of people[554]. Such skills are particularly critical in the current study, where research is conducted from a Western perspective about a phenomenon deeply rooted in the Chinese culture. Interviewers must be able to communicate with diverse people, including monitoring and adjusting their behavior to deal effectively with those from different cultures[555]. Interviewers should also give an *appearance* that credibly represents the research institution and is appropriate to the interview situation[556]. Also, stereotypes in the interviewer's appearance and action can introduce bias. For instance, inflections of voice or prompting with smiles, nods, etc. may encourage or discourage specific replies[557].

Particular attention must be paid to the used *language* and choice of words, which must be understandable for the target group[558]. When conducting research in a foreign language, language barriers need to be overcome[559]. This requires either extensive knowledge of the foreign language or the use of interviewers, interpreters, and translators capable of speaking the foreign as well as the language of the researcher[560]. If such resources are not available, interviews need to be administered in a language that the interviewees might not be in full

[550] Bogner/Menz (2009), pp. 46-48.
[551] Bogner/Menz (2009), pp. 46-48.
[552] Bogner/Menz (2009), pp. 46-48.
[553] Kaiser (2014), p. 6.
[554] Döring et al. (2015), pp. 360-363.
[555] Pruegger/Rogers (1994), pp. 369-371.
[556] Döring et al. (2015), pp. 360-363.
[557] Cooper/Schindler (2014), p. 256.
[558] Döring et al. (2015), pp. 360-363.
[559] Welch/Piekkari (2006), pp. 427-435.
[560] Temple/Young (2004), pp. 167-174.

command. Therefore, interviewers must cautiously translate their information target into appropriate questions in terms of interview context and language. The questions must offer as little ambiguity as possible to minimize the risk of misinterpretation[561]. This drawback is common for research along language and cultural barriers and must be considered when conducting an interview[562]. A different challenge is the *translation* of interviews from a different language than the language in which the subsequent documentation and analysis will be made. It must be decided which translation strategy, literal or nonliteral, should be used to address words or phrases that do not have an exact equivalent in the other language and what meanings and messages do words or phrases carry in one cultural context while not in the other[563]. Decisions about the translation approach should be made based on the overall research design, study objective, type of interview data, and the analysis goals[564]. The interviewing research associate for this study was a German native speaker with excellent English language command. Therefore, English was chiefly used for conducting the interviews and where the interviewees were native German speakers, the German language.

Accurate *documentation* is essential for research traceability and analysis. This includes recording and transcribing interviews or making notes through an interview protocol. These transcripts do not necessarily have to be included in every research report, but at least results from the interviews must be documented in the form of text passages[565]. Recording interviews usually is preferable as the interviewer does not have to make extensive notes during the conversation, which reduces the concentration on the interview itself. Also, it is not required to write comprehensive memory minutes immediately after an interview, which will never reach a quality level comparable to a recording. Furthermore, a recording allows to later quote experts literally without additional verification, for instance, through a second source[566]. A recording might sometimes not be desirable if there is a risk that experts withhold relevant information or respond in a socially desirable manner, such as during interviews about sensitive or confidential topics[567].

Especially when these topics are under investigation, *rapport and trust between interviewer and interviewee* are crucial. For instance, in the current study, interviewees might be reluctant to openly talk about Guanxi structures and mechanisms as it might be perceived as close to corruption and creates the impression that personal benefits are obtained at a social cost[568]. Also, admitting their behavior that could be considered non-compliant or illegal might put interviewees at risk of losing their employment or even facing criminal prosecution. Researching such sensitive topics that involve discreditable or incriminating behavior requires sensitivity to the "confidences and intimacies of others[569]." The interviewer might pose a threat or risk to the interviewees who possibly fear sanctions and exposure[570]. Therefore, rapport must be established[571]. The researcher needs to encourage the respondents to "open up" and share their knowledge[572]. It requires developing a trusting relationship and being seen as someone with

---

[561] Gläser/Laudel (2009), p. 112; Stening/Zhang (2007), pp. 129-131.
[562] Gläser/Laudel (2009), p. 112; Stening/Zhang (2007), pp. 129-131.
[563] Filep (2009), pp. 59-61.
[564] Littig/Pöchhacker (2014), p. 1092.
[565] Aghamanoukjan et al. (2009), p. 432.
[566] Kaiser (2014), pp. 83-86.
[567] Gläser/Laudel (2009), pp. 157-158.
[568] Fan (2002a), pp. 377-379; Luo (2008), p. 188.
[569] Renzetti/Lee (1993), ix; Laine (2000), p. 67.
[570] Laine (2000), p. 67.
[571] Dickson-Swift et al. (2009), pp. 69-70.
[572] Hubbard et al. (2001), p. 134.

whom a research participant is comfortable spending time with and talking to[573]. This can be achieved by explaining to potential participants that their contributions are valuable or by being friendly to them. Also, it is imperative not to "push too hard," which means to accept that a "no" must ultimately mean "no[574]." Additional helpful tactics can be to ensure confidentiality or to show the researcher's reputable institution's approval [575]. Paying attention to confidentiality helps to legitimize the overall research process and encourages respondents' future accurate reporting[576].

Interview Guide design

An important tool for expert interviews is the interview guide[577]. It provides a stable scaffold for the data collection and the data analysis process[578]. The interview guide also ensures comparability of interviews and supports the interviewee to follow its themes[579]. The interview guide can still be adjusted to the interview situation so that also questions are not given verbatim but are flexibly formulated by the interviewers to match the interview situation and the interview partner[580]. The guide also helps assess the required interview time, as experts are typically short on available meeting times, and the interview process must therefore be designed to fit within a specific time frame[581].

The content of an interview guide appears on a superficial level as based on a typical, daily conversation. However, as the interviewers have an information target, they must lead the discussion and ensure through their questions that the interview partners will provide the desired answers[582]. On the other hand, the interviewees must follow the interviewers' signals and requests and provide the required information[583]. Questions and themes in the interview guide are developed out of the information target, which is derived from the research question and, if applicable, theoretical pre-considerations[584]. Questions should be posed openly enough to invite interviewees to give comprehensive replies and emphasize important points[585]. A sufficiently wide range of problems should be addressed so that the respondents have an as large as possible chance to react unexpectedly. Topics and questions should be raised in precise enough form that specific meanings and important details in individual interviewees' statements can be reconstructed[586]. The interview guide should also be designed to allow interviewees to describe situations and their involvement with sufficient depth. Likewise, the personal and social context of behavior and reactions should be satisfactorily considered[587].

Questions are usually sequenced so that biographical, basic information such as age, education, or profession is coming in the beginning to give the interviewer an idea of the interview partner[588]. The actual interview then starts with a question that introduces the topic while not

---

[573] Miller/Tewksbury (2001), p. 55.
[574] Wenger (2002), pp. 265-266. However, Wenger also points out that discussion and negotiation are still considered acceptable, cf. Wenger (2002), pp. 265-266.
[575] Liamputtong (2011), pp. 51-52.
[576] Lee (1993), pp. 164-179.
[577] Meuser/Nagel (1991), p. 448.
[578] Taylor et al. (2016), pp. 122-123.
[579] Taylor et al. (2016), pp. 122-123.
[580] Döring et al. (2015), p. 372.
[581] Bogner et al. (2014), pp. 37-38.
[582] Gläser/Laudel (2009), pp. 111-115.
[583] Gläser/Laudel (2009), pp. 111-115.
[584] Gläser/Laudel (2009), pp. 115-116.
[585] Mey/Mruck (2011), pp. 260-262.
[586] Hopf (1978), pp. 99-100; Gläser/Laudel (2009), pp. 115-116.
[587] Hopf (1978), pp. 99-100; Gläser/Laudel (2009), pp. 115-116.
[588] Döring et al. (2015), p. 372.

being too specific. Such a question should be relatively easy to answer and not overload an interviewee; it should also not be too private or aim at something critical[589]. It is then followed by general questions about the topic under investigation, supplemented by detailed questions. Sensitive, intimate, or critical questions should be raised at the end of the interview when a rapport between interviewer and interviewee was already created[590].

To meet the information target for this study, the designed interview guide was mainly structured around the selection process for LSPs[591]. After the guide's draft was created, following the recommendations of, for example, Döring et al. or Cooper and Schindler, pretests were conducted[592]. A key account manager of a large, international LSP and an employee in the logistics department of a Tier 1 automotive supplier were selected as participants. The pre-test style followed the actual intended interview, and the process and the questions were afterwards reviewed jointly with the interviewees. This helped assess the required time to complete an interview and improve the questions' content and wording. The final, updated version of the interview guide contains seven sections. The first section aims at obtaining the professional background and experiences of the interviewee. Section one, therefore, covers the current professional position, the years of work experience (as a proxy for assessing the quality of the information given), and the participation in selection processes for LSPs. Additional focus is on the Chinese market experience, especially when interviewees are not Chinese citizens or have worked abroad for a more extended period. Afterwards, background questions about the interviewees' current employer are asked. This is to provide insights about which specific points the interview might focus on and identify possible areas for in-depth questions. Sections three to five follow the selection process for LSPs. For each step, questions about the influence of Guanxi as a possible advantage over competitors are posed. In the last section, information about the personal and academic background of the interviewee is asked as an additional proxy to verify the interviewee's expertise from an academic perspective. The complete interview guide is shown in Appendix i1.

5.2.3. Participants

Decisive for type and quality of the information provided through interviews is the selection of the interview partners[593]. In this section, the selection process will be introduced, and the profile of the respondents summarized.

Selection

The selection process must be based on the research question and considering that experts should be found that can provide information about the research topic[594]. Although varying definitions of an expert exist in the literature and the term is controversially discussed, an expert is frequently referred to as a person that holds "specialist craft or knowledge[595]." Based on the understanding that somebody's knowledge is particularly interesting if these persons are also in a position where it is practically applied, several researchers add "power" as another dimension to this view[596]. Experts can then be considered as persons "who have privileged

---

[589] Mey/Mruck (2011), pp. 260-262.
[590] Döring et al. (2015), p. 372.
[591] Pfohl (2016), pp. 345-346.
[592] Döring et al. (2015), pp. 372-373; Cooper/Schindler (2014), pp. 323-325.
[593] Gläser/Laudel (2009), p. 117.
[594] Bogner et al. (2014), pp. 34-35.
[595] Grundmann (2017), p. 26.
[596] Bogner et al. (2014), p. 13.

access to information and – moreover – who can be made responsible for the planning and provision of problem solutions[597]." The following guiding questions should, therefore, be considered when selecting interview partners. Who possesses relevant information? Who has the highest ability to provide accurate information, including who possess sufficient language skills[598]? Who is most likely willing to provide information? Who of the informants is available[599]? Furthermore, as no exact information about the distribution of relevant knowledge and the power structures within a field of study are available in advance, the selection of interview partners must be an iterative process[600]. After the first interviews are conducted, new insights are available, which can help to select further participants[601].

This study aimed to identify the possible impacts of Guanxi on the selection of Logistics Service Providers in China, focusing on the automotive industry. Therefore, participants from buying centers of companies in the automotive industry, i.e., members from logistics and purchasing departments as well as from the selling center of LSPs, i.e., members of the operations and sales departments, in China were targeted[602]. Thanks to the support of a major automotive Tier 1 supplier in China, access to current and former logistics and purchasing associates was granted. The company has some 65,000 employees in China, out of which almost 1,000 are working in a logistics function on central, divisional, and plant level, and around 20 in a logistics purchasing function. The cooperation also led to access to LSPs operating in China that are currently or were previously in business relations with this Tier 1 supplier. Associates working at LSPs often switch positions between operations and sales functions throughout their careers. Therefore, and as maintaining relationships is one of their main tasks, the focus was to identify key account managers with several years of work experience at LSPs. Such key account managers deal with various buying companies and, therefore, typically have deep insights.

Out of around 500 potential interview partners, associates were selected who were in command of the English or German language, presumably had the highest ability to provide accurate information, and were most likely willing to share it. Despite high efforts, none of the associates currently or previously working in the logistics purchasing functions agreed to the interview. This might be related to the sensitive nature of the topic. The purchasing function is frequently at the center of non-compliant behavior, and therefore it could be difficult to openly share information without some form of self-incrimination (cf. section 5.2.2).

To facilitate the later analysis, all participants were asked for permission to record the interview. About half agreed to it, while the other half declined due to concerns about the subject's sensitivity. In the latter cases, notes were taken during the interview and in the end jointly discussed with the interviewee. Many interviewees, especially those who shared sensitive information that could potentially harm their own, respectively, their current or former employer's reputation, asked to remain anonymous.

Respondents

In total, 22 expert interviews were conducted from July to September 2016, with 14 participants working for LSPs and eight working for Tier 1 suppliers. The home country for 19 of them was

[597] Pfadenhauer (2009), p. 83; Meuser/Nagel (1991), pp. 441-470.
[598] Cooper/Schindler (2014), p. 152.
[599] Gorden (1975), pp. 196-197; Gläser/Laudel (2009), p. 117.
[600] Bogner/Menz (2002), pp. 46-47.
[601] Bogner/Menz (2002), pp. 46-47.
[602] Pfohl (2016), pp. 345-347; Anderson et al. (2011), pp. 97-115.

China, where the interviews were conducted in English, and for three of them, Germany, where the interviews were conducted in German. The most common job title was "Key Account Manager" (KAM) with nine experts, followed by "Plant Logistics Manager with five experts. On average, the participants possessed 13 years of work experience, ranging from three to 27 years. Most had Bachelor (12), some Master (6), and few MBA (4) degrees. Table 1 provides an overview of the experts with their assigned code, the company type they work for, their home country, the job title(s), the years of work experience, their highest academic degree, and how the interview was documented.

Table 1: Expert interview participants

Code	Company type	Home country	Job title	Work exp. [years]	Degree	Documentation
A	LSP	China	Key Account Manager	18	Bachelor	Notes
B	LSP	China	Key Account Manager	13	Bachelor	Recording
C	LSP	China	General Manager Sales	14	MBA	Notes
D	LSP	China	Sales Manager and Director of Strategy	24	Bachelor	Notes
E	LSP	China	Key Account Manager	27	MBA	Notes
F	LSP	China	Key Account Manager	12	Master	Recording
G	LSP	China	Key Account Manager	16	Bachelor	Recording
H_1	LSP	China	Key Account Manager	13	Master	Recording
H_2	LSP	China	Key Account Manager	15	Master	Recording
H_3	LSP	China	Key Account Manager	24	Bachelor	Recording
I	LSP	China	Managing Director	12	Master	Recording
J	Tier 1	China	Central Logistics Manager	5	Master	Notes
K	Tier 1	China	Central Logistics Manager	10	Bachelor	Notes
L	Tier 1	China	Central Logistics Manager	4	Master	Notes
M_1	Tier 1	China	Plant Logistics Manager	5	Bachelor	Notes
M_2	Tier 1	China	Plant Logistics Manager	13	Bachelor	Notes
N	Tier 1	China	Plant Logistics Director	20	MBA	Recording
O	Tier 1	China	(Former) Plant Logistics Manager	8	Bachelor	Recording
P	Tier 1	China	(Former) Plant Logistics Manager	5	Bachelor	Recording
Q	LSP	Germany	Key Account Manager	3	Bachelor	Recording
R	LSP	Germany	Automotive Segment Key Account Manager	19	MBA	Notes
S	LSP	Germany	Automotive Segment Key Account Manager	8	Bachelor	Notes

The interviews of H and M were each conducted in groups with several experts answering questions one after the other. As interview H was recorded, a distinction which expert made which statement was possible. Hence, the expert statements in the analysis are denoted separately by an index. For interview M, a recording was not allowed, which made the distinction later on not feasible. Correspondingly, the statements are not separately denoted.

5.2.4. Analysis

In total, around 120 pages of interview transcripts and notes were collected. Interviews conducted in German were translated into English by the interviewer, who was a German native speaker and possessed an excellent command of the English language. A literal translation strategy was employed, and if words with a distinct meaning in the German language were used by interviewees, a close equivalent in English was chosen for the translation. To analyze the collected data, a qualitative data analysis following the approach proposed by Meuser and Nagel was employed[603]. In this interpretive, highly inductive/data-driven analysis method, the explorative target to build hypotheses and theory is followed[604]. The benefits of this approach are that it is holistic and complex, preserves context and background, as well as that mistakes and misinterpretations, are more likely avoided[605]. It is conducted in five steps[606]. 1. Transcription. Thematically relevant passages of the recorded interview are transcribed, with prosodic and paralinguistic elements only being notated scarcely. 2. Paraphrasing. The text is sequenced according to thematic units. 3. Coding. The paraphrased passages are ordered thematically. 4. Thematic comparison. Like in the coding step, thematically comparable passages from different interviews are tied together. 5. Sociological conceptualization. Texts and the terminology of the interviewees are distantly reviewed. The same and different features across interviews are elaborated and categorized, characteristics of the commonly shared experts' knowledge are condensed, and categories are formulated[607]. The results of this qualitative analysis are grouped following the stages of the selection process for Logistics Service Providers, according to Pfohl[608].

5.3. Results and Discussion

This study led to two major findings about the impact of Guanxi on the selection process of LSPs in China. The first one is support possibilities, ways how a buying center can support an LSP along the stages of the selection process; the second one contingency factors, factors that may alter the support possibilities. These will be summarized in section 5.3.2 and subsequently evaluated in detail throughout section 5.3.2 and 5.3.3. This segment ends by stating the limitations of the study and recommending further research work in section 5.3.4.

---

[603] Meuser/Nagel (1991), pp. 455-466.
[604] Döring et al. (2015), p. 599.
[605] Kuckartz (2008), pp. 66-68.
[606] Meuser/Nagel (1991), pp. 455-466.
[607] Meuser/Nagel (2009), pp. 35-36.
[608] Pfohl (2016), pp. 345-346.
[609] Lin/Shao (2000), pp. 283-284. Throughout the interviews, plenty of additional information, such as the impact of relations with customs officials or how Guanxi is built through operations, was provided. This is omitted here as it goes beyond the scope of the study.

5.3.2. Evaluation Guanxi support possibilities

This section explains the Guanxi support possibilities and critically assesses how they relate to the current research state. Selected quotes of expert statements are integrated, while all relevant interview passages can be found in Appendix i2[610].

Buying center shares actual contract prolongation price reduction target

Conducting a tender project requires considerable effort for both the contractee and the contractor[611]. In addition, implementing a new LSP is a cumbersome undertaking for the contractee[612]. Therefore, it might be especially in the interest of associates in the buying center to avoid new RFQs and instead prolong an existing contract[613]. *"RFQs require a lot of energy on both sides. If there is a chance to avoid the RFQ, then it might be better for both sides. 'A little information can help,' meaning that instead of starting a new selection process, one can give information about price and service requirements to the LSP beforehand and the LSP can adjust a little to provide the service needed to keep business and avoid a new RFQ (C)."* Also, Expert R highlights that *"if tendering should be avoided, personal relationships can play a role. For example, if you want to get a 3% price reduction to avoid an RFQ, most LSPs will do it. At [Tier 1 supplier]*

---

[610] The full interview transcripts can be found in the confidential Appendix i3.
[611] Krasnokutskaya/Seim (2011), p. 2657.
[612] Hofmann et al. (2018), pp. 115-123.
[613] Experts C, F, Q and R.

we recently extended the contract (...) this was initiated by the customer. 'I would like to extend because I have thousands of other topics on the table' or any other reasons why he says, 'I want to extend now', even if it is an extension for only a year. Or maybe for two years, but then you have to re-assemble, he has new requirements. 'But I want the reduction', and then you have to see if you can realize the reduction and then you have to somehow gather together to ensure that you get it done (R)." Purchasing associates, which are typically part of the buying center, often get annual price reduction targets that they must meet, and therefore a negotiation with the LSP is required[614]. Knowing the actual contract prolongation price reduction target in such negotiation puts LSPs in a beneficial situation as they essentially know the contractee's reserve price and the alternative to a negotiated agreement[615]. As a result, this might lead to a prolongation price higher than the actual reserve price of the LSP or above market level[616]. This potentially is a suboptimal result with a net loss for the contractee, even considering implementation costs for another LSP and the costs of conducting the tender. While this risk exists, a contract prolongation without the tender process can also be a net gain for a contractee if the negotiated price is lower than the market price plus implementation costs for another LSP and costs related to the tender.

Buying center shares information about upcoming project before official start

The preparation of bidding documents to respond to RFQs requires considerable time and effort for LSPs[617]. Expert G explains that if the *"relationship is good, for example [before] the bidding, they can inform you early for some business. (...) They can tell you 'we have a new opportunity and are you interested' (G)."* The preparation is especially time-consuming when assets like a transportation fleet or warehouses are not owned by the LSP, which is the case for the majority of large LSPs today, and therefore negotiations with subcontractors required[618]. Expert A and D explain that the quality and prices of the tender offer and tender outcome depend on the time given to prepare. Insufficient preparation time is considered a critical failure factor in the participation of tender projects. *"This information can be highly useful as it enables the LSP to prepare offers for this customer for certain lanes in advance [which gives him] an advantage in the negotiations process. If an LSP does not have Guanxi with the customer there is no chance to ask or obtain this information (D)."* Without reasonable time for investigation and collection of relevant details, bids can contain numerous errors. Also, contractees tend to reduce the overall bidding time to have buffers for tender execution and unforeseen events[619]. In the fraud prevention literature, giving a potential vendor a head start for the bid preparation, for instance, by providing specifications upfront, is considered a means of bid-rigging[620]. The exacerbated version is the case when the overall project time is insufficient for developing bids without such head start, effectively limiting the number of potential vendors[621].

---

[614] Giunipero/O'Neal (1988), pp. 38-39.
[615] Kim/Fragale (2005), p. 380; Pinkley et al. (1994), pp. 110-115. For a description of negotiation theory and corresponding strategies, cf. Korobkin (2014), pp. 1-528 and Carraro et al. (2005), pp. 1-53.
[616] Carnevale/Dreu (2006), pp. 55-76.
[617] Krasnokutskaya/Seim (2011), p. 2657.
[618] Krasnokutskaya/Seim (2011), p. 2657; Hofmann/Lampe (2013), pp. 324-332.
[619] Iyer/Jha (2005), pp. 290-292.
[620] Bid-rigging occurs when prices are raised or quality is lowered for goods/services, because businesses that normally should compete (or here conduct a competitive bidding process among several business), secretly conspire. Organization for Economic Cooperation and Development (2009), pp. 1-11.
[621] Kenney/Kirby (1984), p. 352; Wells (2014), pp. 249-253.

Buying center considers bidder as possible bidder

When purchasing decisions are made, selection criteria can be divided into order-winners and qualifiers[622]. Order-winners are the factors that lead to the winning of an order. However, to be even considered as a possible supplier, certain criteria must be met. Such qualifiers can include certification requirements such as ISO 9000, availability, or reputation[623]. Order qualifiers put a company on the same level as their competitors, order-winners above them. Both attributes are necessary for a company to survive and grow[624]. Trust is an important element in buyer-supplier relationships on a personal and on a company level[625]. Especially in China's transitional economy characterized by often underdeveloped institutions, trust among people is generally low[626]. Xinren, deep trust, is one of Guanxi's defining dimensions so that companies in a Guanxi relationship have the confidence that they can believe in another's word[627]. Guanxi also can be translated as "pass the gate and get connected," which echoes that it might be a necessary qualifier to obtain business[628]. Various experts highlight this. For instance, Expert B explains that *"first, we must know the people at the customers, so we have a chance to be invited to a bidding or global tender (...). First you should have the ice broken, let the potential customer know your company. This is the sales part... that's why companies should have a sales. The salespeople should open the door to this company at first. The policy or KPI is the second part, but the sales needs to open the door* (B)." Also, Expert D confirms this, stating that *"Guanxi is good for starting a business. Guanxi in this sense means personal relationships. It is a door opener for new businesses. After the door has been opened, quality of the service and trust between the companies is more important. Relationship is the only way to propose a value proposition. Especially during business lunches this is a possibility. If the customer has lunch with you it is a sign that they like you and offer you the possibility to propose a quotation or solution* (D)." Likewise, previous studies showed that Guanxi could have a significant and positive impact on a company's market share[629].

Buying center shares information if RFQ is for actual project or only for market benchmarking purposes

Market knowledge, the "in-depth understanding of the market conditions in which a purchaser buys his or her products," is considered a key knowledge area for purchasing employees[630]. This can be gained through supply market research, the "systematic gathering, classification, and analysis of data considering all relevant factors that influence the procurement of goods and services to meet present and future company requirements[631]." Supply market research serves as an objective base for key purchasing decisions and can cover materials, goods, and services, suppliers, systems, and procedures[632]. To collect market price information for benchmarking purposes, RFQs can be sent to LSPs. Expert Q explains that without benchmarks, a contractee might not know where his prices are compared to the market level as *"throughout the contract period, (...) the customer was (...) in blind flight* (Q)." However, engaging in a tender requires

[622] Hill (1995), p. 46.
[623] Lyles et al. (2008), pp. 170-171.
[624] Hill (1995), p. 46.
[625] Doney/Cannon (1997), pp. 44-48.
[626] Atuahene-Gima/Li (2002), pp. 62-74.
[627] Kriz/Keating (2010), pp. 308-309.
[628] Lee/Dawes (2005), p. 29.
[629] Peng/Luo (2000), p. 494; Luo et al. (2012), pp. 158-161.
[630] Giunipero (2000), pp. 6-52.
[631] Van Weele (2010), pp. 131-135.
[632] Van Weele (2010), pp. 131-135.

considerable effort for LSPs as they must prepare a bid, obtain information for the bid preparation, and calculate strategies for winning bidding rounds[633]. Informant R quantified this effort with an average of 30,000 to 50,000 USD (R). Besides, informant A explained that LSPs have only limited capacity and cannot respond to all RFQs. Therefore, especially during the peak season for tender projects towards the end of a calendar year, LSPs must select in which tenders to participate (A)[634]. Expert G states that customers *"can also give you information that 'we have to invite you but actually for this lane we will not select you.' They give you some information in advance that you don't need to focus on this if you have very good relationship. 'Because for some lane we use [a certain LSP] for more than 20 years [so] we don't want to change'. Just let you use Germany-Shanghai for 20 years, don't want to change. [They will tell you we have to do the tender process, but] you don't have chance for this lane, so we can focus on others. [Normally] you can't know this before. You have first round, second round, and third round. So, second round they will say 'we will not invite you because you are not potentially to be selected.' So, [depending] on the customer... you know them very long time already so they can tell you before the first round, but if you [are in a] very official [relationship], they can't tell you (...) Depends on how long you know the customer and the company policy. [So we can save time and money] and we can concentrate on where we have most chances to win (G)."*

Buying center evaluates bidder's quality more positively

Expert D explains that *"Guanxi is usually built through operations. Good performance of a company will lead to better Guanxi and better Guanxi will lead to sales in the future. The people from operations can influence the decision made in the headquarter by declaring the quality of the LSP as very high, therefore giving that LSP an advantage in the selection process. As price is mostly not the most important criteria, the quality criteria come into consideration and can be influenced. This leads to internal lobbying (D)."* Also, Expert J validates this phenomenon that *"if a Logistics Service Provider has a good Guanxi with the logistics function they can rely on favorable presentation and judgement. (...) The employees of a company have Guanxi between each other and possibly to management as well. If the management bases their decision on quality reports made by a team leader, and this team leader has good Guanxi with the management, they will select the service provider due to the assessment. This works only if the choice for selection of Logistics Service Providers can be made by a limited circle of people. Interesting to note is that giving a favorable review is of low risk for the employee. An overly positive review of a supplier can always be admitted to past experiences and performance of the supplier. If the supplier doesn't perform according to the expected service, the employee might not be held accountable as the mistake lies with the Logistics Service Provider (J)."* Quality is often the most important criterion when selecting suppliers[635]. In an LSP selection process, quality and reliability are among the most important criteria, far above factors such as relationship and collaboration[636]. Therefore, personal relationships should only have a small impact on the evaluation of an LSP's running business. However, the experts' statements can be explained based on the findings from research about reciprocal commitments in trusted relationships[637]. If related parties evaluate a relationship positive, typically due to expected positive future benefits and the identification with collective goals and values, they might have an "enduring desire" to maintain a valued

---

[633] Whitford (2007), p. 72.
[634] Whitford (2007), p. 72.
[635] Simpson et al. (2002), p. 32.
[636] Alkhatib et al. (2015), p. 133.
[637] Luo (2002), pp. 677-692.

relationship[638]. As Guanxi is increasingly developing among two parties, favors are exchanged with the objective of obtaining benefits. Therefore, especially considering that reciprocity of favor exchanges is seen as the "most pervasive rule guiding Chinese social and economic interactions," it is conceivable that customers evaluate the quality of an LSP, with which they are in a Guanxi relationship and to whom they owe favors, overly positive[639].

Buying center shares information for which tender package bidder has highest chance to win

A common practice for tendering companies is to split large projects into a series of packages[640]. This can be a contractual form of risk control or to account for a fragmented supply market where various providers have different strengths[641]. In addition, when complementarities over goods exist, using a packages-based instead of a traditional bidding procedure can help to reduce a contractee's costs[642]. Such tender approaches are increasingly used by large firms when procuring logistics services[643]. Especially for tendering transportation services, packages of multiple lanes can help to account for an LSP's economies of scope advantage, e.g., to balance transportation networks and increase equipment utilization[644]. Knowing for which tender package a bidder has the highest chance to win allows an LSP to focus its resources. Expert F states that *"you might get hints, don't focus on this part, focus on the other part because we have already decided for this supplier (...) If you have a relationship, they are more prepared to give you the business, maybe because they like you or are comfortable working with you so they give you more information ...like we should focus on this lane. (...) If you have a relationship, coming back to Guanxi in logistics, it comes to information. Knowing where you have to focus. Maybe they are going to invite four providers to beat us, but they really like this one, because this person knows this person, and this person knows that person (F)."* As tender outcome and quality often depend on the preparation intensity, this can, depending on the tender size and number of packages, give an LSP a crucial advantage[645]. Besides, it also increases the risk of bid-pooling, a process where several bidders collaborate to ensure that each wins a certain tender package. LSPs can pre-align for which package which LSP submits a winning bid, while the others deliberately bid at inflated prices[646].

Buying center shares information about competitor's price

"In order to win a negotiation, information regarding the rates of competitors can be useful. The purchasing department of the customer might give you information about the rates of your competitor, but it might be hard to trust this information. False information can be given out as well just to lower prices and to increase competition (expecting better offers). So Guanxi can be a mean to be able to trust this information. Depending on the closeness of the person giving out the information (family, business friend) the information might be more or less trustful (D)." Competitor analysis is fundamental to achieve a competitive advantage[647]. In a study by Ghoshal

---

[638] Moorman et al. (1992), p. 316; Schmoltzi/Wallenburg (2012), p. 56.
[639] Chen/Chen (2004), pp. 317-319.
[640] Drew/Skitmore (1997), p. 471.
[641] Gefen et al. (2008), pp. 531-532; Jiang (2018), pp. 1-24; Hong/Liu (2007), pp. 62-64.
[642] Elmaghraby/Keskinocak (2004), pp. 247-248.
[643] Vries/Vohra (2003), pp. 284-286.
[644] Freight lanes in tender projects are one-way movements from an origin to a destination with the associated set of shipments for the period covered by the RFP. Jara-Diaz (1988), pp. 159-170; Sheffi (2004), pp. 245-247.
[645] Expert A, D; Whitford (2007), p. 72.
[646] Wells (2014), pp. 249-253.
[647] Porter (1998), pp. 361-471.

and Westney, companies consider the understanding of competitors as extremely important and rate its importance as further growing[648]. Competitor intelligence, "identifying, monitoring and understanding specific current competitors," is recognized to have an impact on sales and business development[649]. A study showed that the second most important intelligence topic in this several billion USD market is competitive benchmarking, including the obtainment of estimates of competitors' manufacturing costs, workforce strengths, wage bills, or R&D expenditures[650]. Commercially sensitive information, "information that has economic value or could cause economic harm if known," are trade secrets, financial, commercial, scientific or technical information, as well as "information whose disclosure could prejudice the conduct or outcome of contractual or other negotiations of the person to whom the information relates[651]." A competitor's prices can be considered their trade secret, and sharing them by (potential) customers might violate laws[652]. Depending on the circumstances, communicating prices to a firm's competitor can also constitute a violation of antitrust laws[653]. Receiving information about competitors' prices becomes particularly critical when bids for a tender do not have to be submitted precisely at the same time. LSPs can have a vital advantage when they know their competitors' quoted prices before submitting their bids or if it is allowed to adjust their own bids before the end of the "Offer normalization / adjustment and negotiation" phase. In such cases, LSPs could narrowly undercut competitors and hence secure a contract[654]. This phenomenon is also shared by Expert H_2. *"They have some preferred Logistics Service Provider who they'll go for and sometimes these local service providers are the brother or sister or cousin or relative sitting in the company. So even if you (…) as an international service provider, if you want to penetrate and do some business [it is not easy because] the price we are willing to give, somehow (…) will be released to someone there and they will do match it or they may give you some reason that you will not be able to come into the business because of ABC (H_2)."* Studies also showed that in cases where efforts are needed to gain advantages in the form of a resubmission possibility, for instance, through gifts to the purchaser, bidding prices are inflated, and collusion among bidders is facilitated[655].

Buying center shares information about bidder's and competitor's rank

Expert G highlights that *"if you have good relationship, [a] customer will give you some general, not violate the company policy, but give you some guideline, some guide, indication… (…) [Tell you] can you still lower how many percentages or how much (G)."* Expert L confirms this notion, stating that *"if an LSP has a very good relationship with the customer of course they can get this relevant information. Because for this evaluation information it is confidential. But at least someone will know this, because they need to evaluate the LSP. There is not just one person to do this job. If more than one person to do this job, then there is a chance for the LSP to get this information (L)."* In a competitive tender process, participating LSPs must estimate their costs and add a profit markup. While a lower bid increases the chance of winning a tender, a higher bid could lead to greater profit[656]. Contractees do not have to select the lowest priced bid but

[648] Ghoshal/Westney (1991), pp. 19-20.
[649] Frates/Sharp (2005), pp. 17-18; Calof/Wright (2008), pp. 719-725.
[650] Frates/Sharp (2005), pp. 17-18; Calof/Wright (2008), pp. 719-725; Ghoshal/Westney (1991), p. 25.
[651] Rosenblum/Maples (2009), pp. 33-34. For additional information such as how "commercially sensitive information" differs from "private information" cf. Rosenblum/Maples (2009), pp. 33-34 and Rosenblum/Maples (2009), p. 42.
[652] Bridy (2008), p. 188.
[653] Thompson (1983), p. 939; Posner (1978), pp. 1187-1197.
[654] Wells (2014), pp. 253-254.
[655] Compte et al. (2005), pp. 10-11.
[656] Boughton (1987), pp. 87-94.

instead typically decide based on the price and nonprice features, optimizing for total value received by a potential contractor[657]. Throughout the "Offer presentation and choice of partner" phase of a typical LSP selection process, various rounds of negotiations and offer adjustments between the various bidders and the contractee are happening[658]. Here, bidders struggle to evaluate their own in relation to competitors' ranks, which are based on the offered prices and the client's evaluation of nonprice features. Bids typically depend on bidding strategies for the tender project, which incorporate company objectives such as profit maximization, gaining of market share, asset utilization, or target ROI achievement[659]. Meanwhile, a contractor's evaluation can consist of elements such as on-time performance, service quality, communication, or service speed[660]. LSPs who know their rank relative to their competitors are in a very advantageous situation as they can strategically optimize their bids[661]. Information about participants' ranks in a tender is also valuable for a company's competitor intelligence efforts[662]. Analyzing a competitor's bidding history allows a company to predict its future bidding strategy more accurately and increases the understanding of how potential clients perceive the companies in the market[663].

5.3.3. Evaluation Contingency factors

The following section discusses the contingency factors contractee/contractor characteristics, project characteristics, and personal characteristics, and evaluates how they fit to previous findings. Selected expert quotes are integrated into the discussion. All relevant interview passages can be found in Appendix i2[664].

Contractee / contractor characteristics

<u>Internationalization degree</u>

Several experts explained that the "Internationalization degree" is a factor mediating the impact of Guanxi in a tender process[665]. Expert A states that international companies *"try not to do business with [purely] local companies because of their compliance rules and the need for use of Guanxi. Local companies as well try to avoid doing business with international LSPs as they have strong company compliance and won't offer favors associated with Guanxi. This will probably not change within the next 5 years as local companies will continue relying on local LSPs because of their Guanxi. (…) When it comes to local companies, they still do it the "old" style. Meaning [that] they give and receive cash although they have become more careful in dealing with Guanxi as the anti-corruption campaign has made it harder to stay undetected. (…) Guanxi is the deciding factor. An international LSP might still be able to offer a lower price than the local LSP due to more efficient operations, but the local LSP usually can make use of Guanxi (e.g. by using Hongbao) to win contracts. International LSPs can't and won't play this game with [local] Chinese companies (A)."* Expert C explains that *"there is a separation between local and international customers. For international customers, usually the sales office in their home country (e.g. an American company*

---

[657] Kelly/Coaker (1976), pp. 285-293.
[658] Pfohl (2016), pp. 345-346.
[659] Boughton (1987), pp. 87-94.
[660] Spencer et al. (1994), pp. 70-71.
[661] Wells (2014), pp. 253-254.
[662] Ghoshal/Westney (1991), pp. 19-20.
[663] Boughton (1987), pp. 87-94.
[664] The full interview transcripts can be found in the confidential Appendix i3.
[665] Experts A, B, C, E, F, G, H_1, H_2, H_3, I, J, M, S and Q.

is contacted by the American sales office) is responsible to initiate the business relationship. With local Chinese customers it is hard to have a business opportunity because of the fierce competition on price with Chinese private owned LSPs. The way Guanxi works with Chinese LSPs and Chinese customers is a simple money cycle where cash comes to play the biggest role. Essentially, a deal is negotiated between company A and LSP X, employee a1 who negotiated the deal or who made sure that LSP X got selected will receive a Hongbao with cash or other benefits (e.g. "entertainment" which can be invitation to KTV with "pretty girls"). The cashflow goes A → X → a1 (...) Local (Chinese) companies don't evaluate multiple LSPs, as relationships are more important than prices or costs (C)." Although various attempts were made and the question was considered central for theory and practice, estimating a firm's internationalization degree remains arbitrary[666]. Various authors developed multiple indices to measure a company's activities across countries. These are either based on unidimensional variables, such as sales, assets, employment, profits or R&D, or composites of them[667]. Practitioners often do not distinguish organizational types according to typologies used in international management research, although it is recognized that they increasingly have a shared definition of what a modern international, global, transnational company is[668]. When distinguishing between domestic and such "international" companies as done by the experts, a commonly used definition approach is from the Organization for Economic Cooperation and Development (OECD). The OECD defines them as "companies or other entities established in more than one country and so linked that they may co-ordinate their operations in various ways[669]." There is debate among scholars if the internationalization of businesses will lead to homogeneity, polarization, or hybridization of business practices[670]. This becomes especially controversial when attempting to operationalize implications of internationalization by, for instance, trying to develop internationally valid business codes of ethics[671]. Nevertheless, various international codes of conduct were released by, e.g., the International Chamber of Commerce or the OECD[672]. When analyzing such global codes and comparing them with individual companies' codes of Ethics and the related literature, certain universal values, including fairness and responsibility, can be derived[673]. An increasing number of countries are also passing laws that impact a company's operations internationally. For example, the "US Foreign Corrupt Practices Law" that prohibits American companies from conducting bribery anywhere in the world, or the "Code of the Confederation of Danish Industries," according to which Danish parent companies are being held legally accountable if it can be demonstrated that they have turned a blind eye to the activities of their national subsidiaries[674]. Companies that operate internationally are increasingly subject to such regulations. Some of the practices close to Guanxi might be perceived as corrupt and therefore violate international laws and codes of conduct[675].

<u>Size</u>

According to expert G, small companies *"don't have a KPI or this kind of global [performance] review or lead time [measurement]... if you like you can benefit (...) personally because you don't*

---

[666] Sullivan (1994), p. 325.
[667] Ietto-Gillies (1998), pp. 18-19; Sullivan (1996), pp. 180-191.
[668] Sambharya (1996), pp. 739-742; Young et al. (2016), pp. 63-65; Bartlett/Beamish (2018), pp. 1-4.
[669] Organization for Economic Cooperation and Development (2008), p. 12.
[670] Asgary/Walle (2002), pp. 69-70.
[671] Asgary/Mitschow (2002), pp. 240-245.
[672] Asgary/Walle (2002), pp. 69-70; Getz (1990), pp. 567-577.
[673] Schwartz (2005), pp. 36-39.
[674] Asgary/Mitschow (2002), p. 240; Lindgreen (2004), pp. 35-37.
[675] Zhan (2012), pp. 99-105.

have measurement (…) to see if your service is good or not. It is just from my mouth how I talk. Because they don't have the overview of the whole picture. But for the international company it is a company manner, not an individual. For the small company it is individual, (…) even when a little late, it is ok (G)." This is also supported by Expert R, who stretches that "*medium-sized LSPs focus on smaller firms. There, personal relationships are extremely important, especially when it comes to national or regional deliveries. These are usually firms with 80-200 employees and without a professional logistics purchasing department (R).*" There is no universal agreement what constitutes an SME. For instance, the US government's definition sets an upper limit of 500 employees, the UK's 250[676]. To reduce the various definition approaches on a national level, the Commission of the European communities developed a recommendation for micro, small and medium-sized enterprises based on a company's staff headcount and a financial criterion[677]. According to this EU recommendation, SMEs are "enterprises which employ fewer than 250 persons and which have an annual turnover not exceeding EUR 50 million, and/or an annual balance sheet total not exceeding EUR 43 million[678]." Studies investigated the relationship between non-compliant behavior and firm size. For instance, some show that there is a systematic pattern of bias against SMEs. They are more affected by bureaucracy and property rights issues, including corruption[679]. However, they are also more likely to violate anti-corruption regulations than larger companies[680]. Various reasons for this are hypothesized. As individual transactions have a higher impact on them, SMEs have a higher incentive to conduct closed business deals, especially if such deals can decide whether a company continues to stay in business[681]. Another viewpoint is that it is caused by the higher degree of decision-making power and the number of functions performed by individual associates. Employees enjoy a high degree of autonomy and are rarely subject to checks from their colleagues[682]. Furthermore, it is assumed that the company culture typically prevalent at SMEs, with high trust among individuals as "everyone knows each other by name and face," leads to a lower focus on compliance and overall relaxed defenses against fraud[683].

Project characteristics

<u>Project volume</u>

According to Expert J, Guanxi plays a "*role when managers can freely and without a review decide for one supplier. This is usually true for small parts of logistics services [such] as courier express parcel. In one case [the LSP] was selected as a delivery service [provider] without a review although service and price have been assumed to be worse than at their direct competitors (J).*" Informant L also mentioned project characteristics as important, where decisive information is provided "*if the tender is very big (L).*" Organizations often establish volume thresholds to distinguish which regulations should be applied in procurement projects. For instance, several approvers are needed above a certain volume threshold, and the tender project will be announced to several potential service providers[684]. Contrary, in his study on how multinational enterprises manage political and social forces in an emerging foreign market, Luo revealed a positive correlation

[676] Bacon/Hoque (2005), p. 1981.
[677] European Commission (2003), pp. 36-38.
[678] European Commission (2003), p. 39.
[679] Schiffer/Weder (2001), p. 37; Krasniqi (2007), pp. 77-88.
[680] Wells (2014), pp. 41-42.
[681] Lindgreen (2004), p. 35.
[682] Wells (2014), pp. 41-42; Linnenluecke/Griffiths (2010), pp. 358-360.
[683] Wells (2014), pp. 41-42; Linnenluecke/Griffiths (2010), pp. 358-360.
[684] Monczka (2009), p. 574; Ganuza/Hauk (2004), pp. 1470-1471.

between project size and perceived corruption[685]. Similarly, Ahlin and Pang demonstrated that the effect of corruption is mediated by the size of investment needed in a given industry, whereby industries with larger required investments are more vulnerable to corruption[686]. Tanzi and Davoodi also hypothesized that as "commission fees" given to decision makers to win contracts are relative to the contract volume, projects get artificially enlarged in size and complexity[687].

<u>Number of involved decision makers</u>

Expert M explained that when *"cross-departments are involved in the selection process it is nearly impossible to recommend an LSP (because of Guanxi). LSPs are selected on a criteria catalogue and by price quotes. Relationships might not be the dominating role in it* (M)." Expert S elaborates that *"personal contacts are always good, but at the end of the day, the price decides, and it is not only a buyer who selects, but a buyer plus someone from the business unit... (...) So it is already a multi-eyes principle which leads to the decision for a LSP. Then there is hardly any chance to gain an advantage through personal relationships (...) [The more people are involved in such a process, the more professional is the process and the less influence have personal relationships.] I mean that's the way it is... If you look profanely at the police. Why do policemen always patrol in pairs? This is not just a two-eye principle to protect each other... No, because... If one accepts bribes, yes, then you have to bribe already both policemen, that they overlook the punishment. Will two people shut up? I don't know. And this is similar... That's how I see it in the purchasing process. That several factors always play a role... So, several parties are always there .... so several pairs of eyes take a look. That such a decision is not made by one person alone. (S)"* The literature on minimizing risks for non-compliant behavior such as corruption draws on lessons from crime prevention actions[688]. Here, opportunity-reducing techniques like the "increase of the perceived risks" through some form of surveillance are recommended[689]. A security principle that accounts for this technique is the four-eye principle, a "segregation of the execution of critical tasks and associated privileges among multiple users." Guidelines for preventing and combating corruption in police authorities also suggest following this principle, by, for example, never working alone[690]. For the purchasing function, that could mean that purchase order approvals should require two signatures, or that the person who issues a purchasing request should be different from the person who approves it[691]. Generally, to manage business processes, companies employ workflow management systems that enable a business to "analyze, simulate, design, enact, control, and monitor its business processes." Recommendations for designing such systems also include the implementation of authorizations that enforce separation of duty constraints to prevent fraud[692].

<u>Usage of structured selection process</u>

---

[685] Luo (2006b), pp. 755-762.
[686] Ahlin/Pang (2008), p. 427.
[687] Tanzi/Davoodi (1998), pp. 43-44.
[688] Gorta (1998), pp. 67-87.
[689] Clarke (1997), pp. 18-21. Clarke also distinguishes between formal surveillance (e.g. security personnel), surveillance by employees (park attendants, CCTV systems) or natural surveillance (e.g. people that "naturally trim bushes at the front of their homes" or "neighborhood watches"), cf. Clarke (1997), pp. 18-21.
[690] Poerting/Vahlenkamp (1998), pp. 243-244.
[691] Meersman et al. (2014), p. 243.
[692] Liu et al. (2004), pp. 375-376.

"It is important to note that mostly only international clients have a defined process (...) Within this process, Guanxi is hard to play a role as this would make this process obsolete. It seems rather complicated to gain an advantage, at least an advantage that is not back-traceable. Important to note is that when there is no RFQ, meaning no official pricing quotation process, then Guanxi can play a major role. Prices will be negotiated with a certain decider within the customer and this opens up a big window for Guanxi to get used (C)," Expert C explained. Structured processes are "formally defined, standardized processes [693]." Characteristics are that they have a predetermined input, an ex-ante specified output and that they can be identically repeated[694]. Standardized processes can facilitate communication, decrease process errors, and allow for incorporating expert knowledge into decisions[695]. The risk of unintended influences increases when tasks are unique[696]. This is the case for the purchasing of services, as services cannot be duplicated exactly for each customer, and therefore defining requirements and measuring quality standards becomes a challenge[697]. In addition, standards that are defined but not followed can be an indicator of harmful influences[698]. The importance of processes is also considered a critical element in the fraud prevention literature. The mechanism most associated with reducing fraud costs is the management review of internal processes, together with a review of internal controls and accounts[699].

Personal characteristics

<u>Age</u>

Changes in personal values across generations can be observed all over the world[700]. Especially changes in the economic environment were typically accompanied by changes in the values of younger generations[701]. For instance, through China's rapid economic development and opening up, personal values shifted towards more individualism[702]. Some scholars consider the phenomenon of Guanxi and its related values as dynamic, and some studies show that younger generations regard Guanxi as "old-fashioned[703]." Younger people believe that resorting to Guanxi at work creates an incompetent, negative impression and, hence, overall employ it less, for example, when looking for jobs[704]. Also, expert J highlights that *"Guanxi is prevalent in the older generation of Chinese. They make use of it a lot and it plays an important role. Due to the shared culture, the younger generation has to deal with Guanxi as well, although they regard it as partly unfair. They try to accomplish their work without the use of Guanxi (J)."* This is seen similarly by expert S who has *"the feeling that maybe... the slightly older generation ... still did not really get there (S),"* to a Guanxi free business environment. On the other hand, for the young generation *"all [is] official. Not that you sit together in the evening and say, OK, you have to get much better at this price or you have to get better at that price (S)."* Scholars in the field of

---

[693] Kroenke/Boyle (2017), p. 286.

[694] Schäfermeyer et al. (2010), p. 2.

[695] Toni/Panizzolo (1993), pp. 1379-1381; Wüllenweber et al. (2009), pp. 527-548; Manrodt/Vitasek (2004), pp. 1-23.

[696] Anand et al. (2004), pp. 49-50.

[697] Jackson et al. (1995), pp. 100-106; Fitzsimmons et al. (1998), pp. 370-379; Pleger Bebko (2000), pp. 9-10; Ellram et al. (2007), p. 47; Wynstra et al. (2018), pp. 90-92.

[698] Labuschagne/Els (2006), pp. 42-43.

[699] Wells (2014), pp. 42-48.

[700] Inglehart (2008), p. 130; Loughlin/Barling (2001), p. 543; Thornton/Young-DeMarco (2001), p. 1009.

[701] Flanagan (1982), pp. 434-436.

[702] Park et al. (2006), pp. 127-132; Zeng/Greenfield (2015), p. 53.

[703] Lovett et al. (1999), p. 237; Luo et al. (2012), pp. 148-152; Chen et al. (2013), pp. 193-194; Wilson/Brennan (2010), pp. 662-663.

[704] Faure/Fang (2008), p. 197; Hanser (2002), pp. 160-161.

crime prevention showed that there is a correlation between age and fraud. Arrests for fraud peak at the age of under 30 and decline linearly, down to less than half, by 41. However, with increased age, the severity and cost of committed fraud increases[705].

<u>Sex</u>

Expert Q points out that *"I also know customers, not from us, but from colleagues of mine, where we say among us, with this guy we've never had business, because we do not go to the brothel with him. Most probably that is not even an exaggeration, but there's another guy who does something once in half a year with him. And of course, there are people or customers, those old men who are susceptible to something like that* (Q)." Various scholars argue that different from the contemporary Western understanding, gender definitions in the Chinese society are considered primal forces that fundamentally shape the world and cannot be separated from social roles[706]. Well into the twentieth century, a woman's place was the "inner world," consisting of family and the household, and a man's the "outer world," consisting of labor and public affairs[707]. Public policies often support the image of women as socially inferior "second-class workers" who rather should stay at home as wives and mothers or conduct work that is deemed suitable for them[708]. The business culture in China is, therefore, traditionally male-dominated, and many Guanxi rituals are built on "sharing the pleasures of masculine heterosexuality" during the night where the provisioning of women's bodies and sexual services is used as a gift[709]. Women are excluded from such activities and are considered to have a different approach to relationships[710].

<u>Home country</u>

Business practices are recognized to vary around the world[711]. Despite the diffusion of certain organizational practices across national borders, such as the spread of small-group activities in Japan, Sweden, and the USA, or the adoption of US human resource practices by European firms, substantial differences in managerial structures and strategies across countries prevail[712]. This can also be traced back to the managerial ideology, "major beliefs and values expressed by top managers that provide organizational members with a frame of reference for action," that influences managers by imposing taken-for-granted world views and creating normative expectations[713]. Such beliefs and values can vary widely and are strongly determined by the national culture of managers[714]. Culture can be defined through its elements, "patterns of and for behavior," whereby the essential core of culture consists of "traditional ideas and especially their attached values[715]." The most common approach to distinguish national cultures was developed by Hofstede, who classifies cultures along the dimensions power distance,

---

[705] Watson (2003), pp. 49-50; Akers (2009), pp. 322-372.
[706] Fei (1992), pp. 26-27; Stacey (1983), pp. 1-324; Anagnost (1989), pp. 313-342; Hershatter (2004), pp. 1013-1017. For instance, according to the concept of yin (Chinese "阴") and yang ("阳") in the Chinese philosophy, each yin and yang are relative concepts "growing out of each other," cf. Smith (1994), pp. 100-108.
[707] Rofel (1999), pp. 1-212.
[708] Robinson (1985), pp. 53-57; Jiang, Jue (2019), pp. 257-267.
[709] Yang (2002), p. 466.
[710] Beetles/Crane (2005), pp. 231-233.
[711] Asgary/Walle (2002), pp. 69-70; Getz (1990), pp. 567-577.
[712] Aguilera/Cuervo-Cazurra (2004), pp. 415-420; Cole (1985), p. 560; Gooderham et al. (1999), pp. 521-528.
[713] Aguilera/Jackson (2003), pp. 457-462.
[714] Laurent (1983), pp. 77-95; Hofstede (1984), pp. 22-26.
[715] Kroeber/Kluckhohn (1952), p. 181. For an overview about various definition approaches and the multiplicity of current views of the term "culture" cf. Jahoda (2012), pp. 289-300.

68

collectivism versus individualism, femininity versus masculinity, and uncertainty avoidance[716]. He also observed cultural differences in the workplace. For example, in rather collectivist countries like China, the relationship between employer and employee is seen as a family relationship where an employee's poor performance is no reason for dismissal as "one does not fire one's child." In contrast, in relatively individualist countries such as the US, the relationship between an employer and employee is mainly a business transaction where the employee's poor performance is considered a legitimate and socially acceptable reason for termination[717]. Expert F shares an *"example of a Chinese customer, [where you know] that there is an RFI coming in one years' time, you have one year to build up relationship, help them. We're a supplier, we should get paid for what we do, but if we can help that customer in terms of benchmark pricing, offering solutions, sharing market knowledge… actually helping this person, maybe that one will be more willing to support you during an RFI period. (…) European approach is very objective, just you know… here is the RFI and we will score it, relationship is still important, obviously you have a better chance if you know this customer for six years than when you know him for six days (…) If you have performance, in theory is measured by facts, which in turn turns into KPIs. I would say with a Chinese customer, if you have good Guanxi then maybe the KPIs are not so strict on facts. If you have a European customer, more often than not, the facts are the facts even if you have a ten-year relationship and a great relationship. With a Chinese customer there are the facts, but maybe more if you have a good relationship what gets reported in the KPIs can maybe be better for the forwarder. I don't want to say it's less honest or real, but sometimes the problems are smoothed over. Similar to what I said about the claims, if you have a good relationship maybe the KPIs look a bit better and there are less claims (F)."*

5.3.4. Limitations and further research

Following is a brief discussion about the limitations of the findings and a recommendation for further research. The manifold implications of this investigation are then discussed in section 6.4 after the subsequent quantitative study.

This study has generated fascinating insights into the possible impacts of Guanxi on the selection of Logistics Service Providers in China. In contrast to previous studies that took a macro view on Guanxi practices in supply chain management, this study could add to the understanding of the phenomenon on a micro level. While these detailed, sensitive findings could only be gained by employing an explorative method, this research approach is also subject to certain limitations. The experts were chosen deliberately based on a presumed ability to provide information as well as assumed motivation to share insights and language skills. While this was done cautiously and a substantial number of experts were interviewed, it cannot be ensured that in case other experts were selected, the same findings would be generated[718]. This becomes especially relevant as the research results cannot be tested for statistical significance. As data was self-reported, there is also a risk of inherent biases such as selective memory or exaggeration[719]. Furthermore, as this study was based on BSRs in the automotive industry, the results cannot be easily extended to other industries. It is conceivable that different phenomena occur in other industries, e.g., which have a higher share of logistics cost in their final product price. In addition, possible SP were described, but frequencies about how often they actually occur cannot be assigned. It is imaginable that some only exist in selected organizations instead

---

[716] Hofstede (2001), p. 31. A review about further cultural patterns is provided in Straub et al. (2002), pp. 14-18.
[717] Hofstede (2001), 119-123.
[718] Kaiser (2014), p. 6.
[719] Brutus et al. (2013), pp. 56-72; Berekoven et al. (2009), p. 97. 69

of being widely spread. While this is a challenge prevalent to most exploratory research designs, it makes an abstraction to theory questionable.

Based on the limitations, further research is required. Quantitative evidence is needed to determine frequencies of occurrence, which would allow for a generalization of the findings. More research on this topic should also investigate whether these are phenomena only occurring in China or common in other countries. Besides, as the target of this study was to identify how Guanxi impacts the LSP selection in China, information from the expert interviews was extracted within this scope. However, the experts also mentioned other points, such as the impact of relations with customs officials or how Guanxi is built during daily operations. An analysis of the interview protocols with a different research aim could help to generate further insights.

6. Survey to verify the impact of Guanxi on the LSP selection in China

The previously described explorative interviews have revealed the possible impacts of Guanxi on the LSP selection in China. To verify the results, as illustrated in Figure 7, these findings will be used as the base for a survey study.

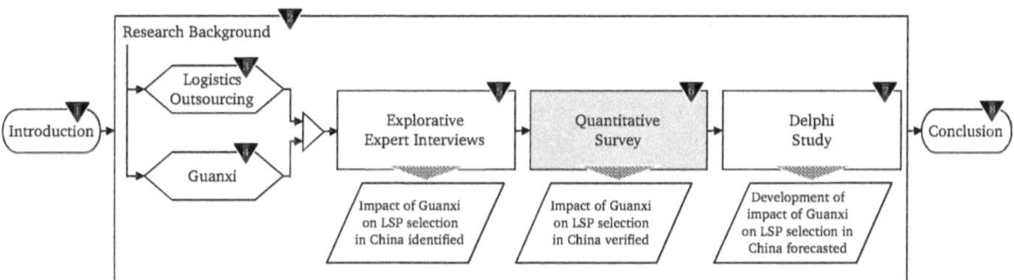

Figure 7: Chapter "Quantitative Survey" within overall research process
Source: Own figure

To that, in section 6.1, the background of this research is explained, and the research questions RQs1, which contingency factors influence the impact of personal relationships on the LSP selection, and RQs2, how do members of a buying center support a certain LSP during the selection process, together with related research propositions will be developed. The employed research approach to answer them is explained in section 6.2. Section 6.2.1 introduces the method and assesses it. Then in section 6.2.2, the questionnaire design and the way how the survey was conducted are explained. It was filled in by 194 respondents, which are, together with the sampling method, described in section 6.2.3. The different analyses techniques that were used are afterwards shown in section 6.2.4. Section 6.3 draws the outcomes together and presents the results, subdivided into sections for the frequencies (6.3.1), crosstabs (6.3.2), Mann-Whitney-U-Test (6.3.3), explorative factor analysis (6.3.4), and regression analysis (6.3.5). The final section contextualizes the research by summarizing the findings in section 6.4.1, evaluating them in section 6.4.2, and indicating limitations and implications in section 6.4.3. This chapter then closes with a recommendation for future research.

6.1. The impact of Guanxi on the selection of LSPs in China

This section will review the background of this study and develop the research questions and propositions.

Background

The previous study identified various support possibilities of Guanxi along the different stages of the selection process and contingency factors that may alter SPs. During the "Project start and acquisition" phase, the buying center can share the actual contract prolongation price reduction target or information about the upcoming project before the official start. In the "Bidder search and RFI" phase, the buying center can consider a bidder as a possible bidder. In the "Preparation of RFPs and offer" phase, the buying center might share information if the RFQ is for an actual project or only for market benchmarking purposes. In the "Offer normalization

/ adjustment and negotiation" phase, the buying center could evaluate a bidder's quality more positively, share information for which tender package the bidder has the highest chance to win or share information about competitors' prices. During the "Offer presentation and choice of partner" phase, the buying center could share information about the bidder's and competitors' ranks. Contingency factors for each contractee and contractor are internationalization degree as well as size. Identified contingency factors related to project characteristics are the project volume, the number of involved decision makers, and whether a structured selection process is used. Personal characteristics that may alter the support possibilities are age, sex, and home country (cf. section 5.3).

While the previous study illuminated Guanxi's impact on the selection of Logistics Service Providers in China, it fell short in describing how common certain practices actually are and whether they are shared around the world or unique to China. This study aims to address this gap and verify the previously found qualitative results quantitatively. For this purpose, a survey was conducted among 194 experts working at LSPs or in supply chain management functions for buying companies. To be able to identify differences between China and other societies, notably the Western one, the survey was distributed globally. Based on the findings from the exploratory study, two main research questions arise.

Research questions and propositions

Studies about the impact of Guanxi led to partly inconsistent results. For instance, while some showed that Guanxi with government officials can lead to increased profits, other studies demonstrate the opposite effect[720]. An approach to explain such inconsistencies is the influence of contextual factors that mitigate the effect of Guanxi[721]. For example, Shou et al. showed that Foreign-controlled businesses are less able to leverage Guanxi networks to enhance performance than their local counterparts[722]. According to Luo and Chen, the link between Guanxi and performance is moderated by the country of investment origin and the entry mode, where for instance, joint ventures are considered beneficial[723]. Moreover, Luo et al. demonstrated that the impact of Guanxi on organizational performance is weaker for privately-owned enterprises than for state-owned enterprises[724]. Similarly, studies about corruption and fraud assume that certain contextual factors impact the prevalence of illicit behavior[725]. Also, in the previously conducted study, various possible contingency factors were identified that might mitigate the impact of Guanxi. To quantitatively verify or falsify them, research question RQs1 states[726]:

> *RQs1: Which contingency factors influence the impact of personal relationships on the LSP selection?*

The identified contingency factors can be distinguished along the three categories "Personal characteristics," "Firm (Contractee/Contractor) characteristics" and "Project characteristics." *Personal characteristics* include the demographic features age, sex, and home country of the individual boundary spanners. The characteristic age can be distinguished in "younger" and "older," sex in "female" and "male," and home country according to the United Nations Member

---

[720] Luo et al. (2012), p. 148.

[721] Adler/Kwon (2002), p. 35; Xiao/Tsui (2007), pp. 23-27.

[722] Shou et al. (2014), pp. 80-81; Ralston et al. (2006), pp. 826-827.

[723] Luo/Chen (1997), p. 14.

[724] Luo et al. (2012), pp. 148-161.

[725] Donchev/Ujhelyi (2014), pp. 309-331.

[726] Due to the international scope of this study, the term "personal relationships" instead of "Guanxi" is used.

States list[727]. *Firm characteristics* include internationalization degree and size of the firm, as well as the firm type – whether the firm is a buying firm (contractee) or an LSP (contractor). Regarding internationalization degree, firms can be classified as "international companies" if they are "established in more than one country and so linked that they may co-ordinate their operations in various ways" and "local companies," if they are not[728]. Firm size can be distinguished in "small," when the company has fewer than 250 people together with an annual turnover of 50 million EUR or less, or an annual balance sheet total of 43 million EUR or less, and "large/big," when it is above either threshold[729]. The firm type can be discerned based on the main commercial activity in "Buyer" of logistics services, i.e., buying firm (contractee), or "LSP," provider of logistics services (contractor). *Project characteristics* can be distinguished in "project volume," "number of involved decision makers" and whether a "structured selection process" is used. Project volume can be divided into "small" and "large" and number of involved decision makers in "few" and "many." The structuredness of the selection process ranges from "unstructured" to "structured." The opposing tendencies of these project characteristics cannot be precisely defined as they depend on a company's size, and therefore individual interpretations have to be made based on the specific company situation. Based on previous research as well as the preliminary findings from the qualitative study about CFs, the following research propositions are made.

The literature has investigated the impact of age and the correlated factor "work experience" on decision-making processes[730]. Different generations possess a different set of values, often shaped by historical and social changes, that reflects on their behavior[731]. Several lines of evidence suggest that younger generations are becoming more individualistic than older ones[732]. They consider resorting to personal relationships in a work setting as outdated and believe that it creates a negative, incompetent impression[733]. Accordingly, the influence of personal relationships on the LSP selection process should be higher for older than for younger decision makers. Therefore, research proposition RPs1 states:

> *RPs1: If the decision maker is younger, the influence of personal relationships on the LSP selection process is lower than if the decision maker is older.*

The participation of women in the labor force has been increasing rapidly over the past decades, leading to a rate of over 50% in OECD countries[734]. In the US, the difference between women's and men's share in the workforce is expected to be less than 10 percent by 2020[735]. However, whether women's patterns of work behavior are equal to those of men has been subject to intense debate[736]. While differences between men and women along biological, neurological, and psychological dimensions are generally recognized, some scholars suggest that men and women are more similar than different[737]. Following this notion, research proposition RPs2 states:

---

[727] Kogan/Shelton (1962), pp. 93-111; United Nations (2020)
[728] Organization for Economic Cooperation and Development (2008), p. 12.
[729] European Commission (2005).
[730] Johnson (1990), p. 75; Dror et al. (1998), p. 67; Bruine de Bruin et al. (2012), p. 352.
[731] Inglehart (2008), p. 130; Loughlin/Barling (2001), p. 543; Thornton/Young-DeMarco (2001), p. 1009.
[732] Sun/Wang (2010), pp. 78-79; Faure/Fang (2008), p. 197; Hanser (2002), pp. 160-161; Chen et al. (2013), pp. 193-194; Wilson/Brennan (2010), pp. 662-663.
[733] Chen et al. (2013), pp. 193-194; Wilson/Brennan (2010), pp. 662-663.
[734] International Labour Organization (2019).
[735] Toossi/Morisi (2017).
[736] Eagly/Johannesen-Schmidt (2001), pp. 781-782; Heilman (2012), pp. 123-131.
[737] Hopkins et al. (2008), pp. 348-349.

RPs2: The sex of the boundary spanners has no influence on the impact of personal relationships on the LSP selection process.

It is now understood that interpersonal relationships and relationships between individuals and groups differ across cultures, based on various factors such as education, industrial development, and learned business practices[738]. As their culture is considered rather collectivist, Chinese people tend to place group goals and collective action ahead of self-interest. On the other hand, Western cultures tend to be more individualistic, so that an individual's self-interest is placed before the group[739]. Accordingly, also business relationships in China differ from the Western ones[740]. They are also social interactions that involve an emotional attachment as well as the exchange of favors and gifts[741]. It has also been observed that business relationships are rather between individuals instead of between functions of organizations[742]. Considering this stronger personal bond beyond organizational boundaries, proposition RPs3 states:

RPs3: The influence of personal relationships on the LSP selection process is lower in Western cultures compared to China

Numerous studies have reported that smaller firms benefit from a higher degree of flexibility, entrepreneurial drive and can rather follow unorthodox ventures[743]. Employees typically enjoy a high degree of autonomy, and the company culture prevalent in SMEs is coined by high trust among individuals, which results in a lower degree of focus on compliant behavior[744]. Evidence suggested that personal relationships are especially for SMEs a source of competitive advantage[745]. Therefore, research proposition RPs4 states:

RPs4: At large companies, the influence of personal relationships on the LSP selection process is lower than in small and medium-sized companies.

Multinational companies are loosely described as firms that are engaging in international operations[746]. They are exposed to various formal rules, e.g., constitutions and laws, and informal constraints such as customs and traditions[747]. Companies accommodate to such institutional arrangements to reap the benefits that international endeavors can bring[748]. For instance, when companies start to conduct business internationally, they increase their CSR activities and the transparency of their financial reporting[749]. Although there are no universally agreed business principles, a growing number of countries accept certain good governance principles such as fairness or responsibility as a base for their laws and regulations[750]. Some of the identified Guanxi support possibilities likely violate such business principles[751]. Therefore, research proposition RPs5 states:

---

[738] Hofstede et al. (2010), pp. 29-32.

[739] Jia et al. (2016), pp. 1249-1250.

[740] Wang (2007), pp. 81-85.

[741] Chen/Chen (2004), p. 314; Kipnis (1997), p. 27; Lee et al. (2018), pp. 357-358.

[742] Davies et al. (1995), pp. 208-213.

[743] Nooteboom (1993), pp. 283-294.

[744] Wells (2014), pp. 41-42; Linnenluecke/Griffiths (2010), pp. 358-360.

[745] Wiegel/Bamford (2015), p. 324.

[746] Sambharya (1996), pp. 739-742; Young et al. (2016), pp. 63-65.

[747] North (1991), pp. 97-111; Keefer/Knack (1997), pp. 591-594.

[748] Hitt et al. (1997), pp. 781-794; Hall/Soskice (2001), pp. 9-27; Carpenter et al. (1998), pp. 158-159.

[749] Gugler/Shi (2009), pp. 3-24; Khanna et al. (2004), pp. 503-504.

[750] Asgary/Mitschow (2002), pp. 240-245; Schwartz (2005), pp. 36-39; He/Cui (2012), pp. 352-355.

[751] Gao et al. (2012), p. 464; Li (2011), p. 20; Braendle et al. (2005), pp. 399-402.

RPs5: At multinational companies, the influence of personal relationships on the LSP selection process is lower than in domestic companies.

It can be assumed that perceptions about personal relationships differ between members of the buying and selling center. Members of the selling center have boundary spanning tasks with incentives to set up close customer relationships as they can differentiate a seller's offer[752]. By contrast, members of the buying center, such as purchasing associates, are often subject to strict compliance rules which discourage personal relationships with suppliers. Ethical codes of behavior often demand to avoid "any personal business or professional activity that would create a conflict between personal interests and the interests of the employer[753]." It is suggested that any activity should be avoided that may be merely perceived as a conflict of interest[754]. Previous research suggests that incentives shape behavior so that the selling center members are likely to set up close personal relationships[755]. Bem and McConnell proposed that the attitude of a person depends on the own behavior and stimulus context in which it occurs[756]. Considering this setting and using the company type where boundary spanners work for as proxy, research proposition RPs6 states:

RPs6: LSPs perceive the influence of personal relationships on the LSP selection process higher than buying firms.

Approvals, authorizations, and verifications are recognized control activities that help to ensure an appropriate risk response of management[757]. Regarding purchasing procedures, it is suggested that additional actions should be adopted once the size of a potential purchase exceeds certain values[758]. Therefore, companies often implement volume thresholds as one criterion to decide which extended regulation, such as additional approvers or an increase in the minimum number of quoting suppliers, should be applied[759]. These measures are intended to avoid risks, such as illicit interferences of boundary spanners. Therefore, research proposition RPs7 is derived as:

RPs7: In case of a large project volume, the influence of personal relationships on the LSP selection process is lower than in case of a small project volume.

A technique to prevent crime is to increase the perceived risk of detection[760]. Because of this, companies adopt "four-eye principles," where the execution of critical tasks and associated privileges are divided among multiple users[761]. To obstruct procurement fraud, duties are segregated in at least two-person systems. For instance, only a different associate is authorized to amend a buyer's list of approved suppliers[762]. Acknowledging that some of the identified

---

[752] Millman/Wilson (1995), p. 13.
[753] Handfield/Baumer (2006), p. 42; Carson (1994), p. 387.
[754] Victor/Cullen (1988), pp. 101-105; Handfield/Baumer (2006), p. 42
[755] Millman/Wilson (1995), p. 13. For a comprehensive discussion about when and why incentives do and do not work to modify behavior cf. Gneezy et al. (2011), pp. 191-210.
[756] Bem/McConnell (1970), pp. 23-30; Killeen (1985), pp. 407-417; Murphy (2000), pp. 245-276.
[757] Moeller (2007), pp. 63-66.
[758] Durant (2005), p. 30.
[759] Monczka (2009), p. 574.
[760] Clarke (1997), pp. 18-21.
[761] Meersman et al. (2014), p. 243.
[762] Murray (2014), p. 8.

support possibilities during the LSP selection process might be related to fraudulent activities, research proposition RPs8 states[763]:

> *RPs8: If a large number of decision makers is involved, the influence of personal relationships on the LSP selection process is lower than if only a few decision makers are involved*

Structured, formally defined, and standardized processes are characterized by a predetermined input, an ex-ante specified output, and that they can be identically repeated[764]. Risks for unintended influences increase when tasks are unique, and therefore it is difficult to define and measure related standards and procedures[765]. Standards that are defined but not followed are a sign of harmful influences, and improving processes is a key mechanism associated with reducing fraud risks[766]. Therefore, research proposition RPs9 states:

> *RPs9: If a highly structured LSP selection process is used, the influence of personal relationships on the LSP selection process is lower than if an unstructured LSP selection process is used.*

As stated earlier, asides from the contingency factors, the qualitative study uncovered eight Guanxi support possibilities during an LSP selection process, all from the perspective of buying center representatives. Although the SPs were identified, it is not understood to what extent they actually occur in practice. This study intends to close this gap, and therefore, research question RQs2 is formulated as follows:

> *RQs2: How do members of a Buying center support a certain LSP during the selection process?*

The individual support possibilities are:

> *SP1: Buying center shares actual contract prolongation price reduction target*

> *SP2: Buying center shares information about upcoming project before official start*

> *SP3: Buying center considers bidder as possible bidder*

> *SP4: Buying center shares information if RFQ is for actual project or only for market benchmarking purposes*

> *SP5: Buying center evaluates bidder's quality more positively*

> *SP6: Buying center shares information for which tender package bidder has highest chance to win*

> *SP7: Buying center shares information about competitor's price*

> *SP8: Buying center shares information about bidder's and competitor's rank*

A survey instrument was developed and administered among supply chain management experts to address the research questions and test the propositions.

---

[763] Zhan (2012), pp. 99-105.
[764] Kroenke/Boyle (2017), p. 286; Schäfermeyer et al. (2010), p. 2.
[765] Anand et al. (2004), pp. 49-50.
[766] Labuschagne/Els (2006), pp. 42-43; Wells (2014), pp. 47-48.

6.2. Method: Survey

The first part of this section will contextualize the survey method and evaluate it. Then, the study procedure, as well as study participants, will be introduced. In the end, a brief synopsis of the analysis approach is provided.

6.2.1. Background and evaluation

Surveys are a versatile method for collecting information through a structured process. Abstract information of all types can be collected by questioning others, which would take much more time and effort to gather through alternative research methods[767]. In standardized surveys, the order and content of questions are fixed and standardized. This standardization ensures that the collected information is comparable across a chosen sample's subsets and can be further processed and evaluated[768]. When combined with certain statistical sampling methods to choose participants, survey findings and conclusions can be projected to large and diverse populations[769].

At the core of a survey is the interaction between a person, a questionnaire, and possibly another administering person[770]. A questionnaire is a more or less standardized compilation of questions that are given to people to answer with the aim of using the answers to verify the questions' underlying theoretical concepts and relationships. This makes questionnaires a central link between theory and analysis [771]. Questionnaires usually measure impressions, attitudes, opinions, feelings, thoughts, or personal data. Typically, there are neither time limits for filling in a questionnaire nor predefined right or wrong answers[772]. Survey communication can be either done in person, assisted via computer respectively telephone, or self-administered, which is paper-based or online. Mostly online and telephone surveys have been gaining popularity over the past years at the expense of other forms[773]. Main advantages of self-administered surveys, especially when computer-delivered, are that they are often the lowest-cost option and independent from time and space. This allows for an expansion of the geographic coverage at low cost and makes them appear more anonymous[774]. Participants also can have more time to think about questions so that more complex questionnaires, possibly including visuals, can be used. Drawbacks are that the response rate is often low, mailing lists have to be accurate, software instructions might be needed, and IT security-related risks[775]. In contrast, telephone surveys' benefits are that the completion time is typically the fastest, and access to hard-to-reach participants can be gained through repeated callbacks. Random dialing via computers can be used, and the results are very current. Like with self-administered surveys, costs are also lower than personal interviews, and geographic coverage can be increased without a dramatic cost increase[776]. Disadvantages of telephone surveys are that response rates are lower than for personal interviews, interview lengths must be limited, and directory listings must be considered unreliable as frequently, phone numbers are unlisted or not working. Also, some target groups might not be available through phone, illustrations cannot be used, and the risk

---

[767] Cooper/Schindler (2014), pp. 218-222.
[768] Berekoven et al. (2009), pp. 93-94.
[769] Cooper/Schindler (2014), pp. 218-222.
[770] Cooper/Schindler (2014), pp. 218-222.
[771] Porst (1996), p. 737.
[772] Schäfer (2016), pp. 29-30.
[773] Reinecke (2019), pp. 725-728.
[774] Cooper/Schindler (2014), p. 225; Wagner-Schelewsky/Hering (2019), pp. 788-789.
[775] Cooper/Schindler (2014), p. 225; Wagner-Schelewsky/Hering (2019), pp. 788-789.
[776] Cooper/Schindler (2014), p. 225; Hüfken (2019), pp. 760-763. 77

for bias through social desirability is higher than with self-administered surveys[777]. To achieve the research goal of this study, data from various regions must be collected. Also, the study subject is sensitive, and time and budget constraints exist. Therefore, an online survey using a questionnaire is considered the most suitable data collection approach.

6.2.2. Procedure

This section will introduce the questionnaire design, considering question-wording, sequencing, data types, and scales, and provide a summary of how the survey was conducted.

Questionnaire design

When designing questionnaires, especially the three fundamental challenges question wording, question sequencing, and data types and scales must be considered[778].

Regarding *question-wording*, it has to be ensured that a questioned person can comprehend the posed questions, the terms used, and grasp the question's meaning. Semantic comprehension, i.e., what does a question or a term in a question mean, and pragmatic comprehension, what does the researcher actually want to know, can be distinguished[779]. Therefore, questions should be stated in a shared vocabulary, with words that have a single meaning and are not biased. Unsupported or misleading assumptions should be avoided, and questions correctly personalized [780]. Furthermore, questionnaires should not exceed a certain length, as participation motivation decreases with increasing time needed to fill in a survey. The acceptable time depends heavily on the surveyed group, the survey topic, and the answering setting so that no standard recommendation can be made[781]. Good practice shows that most questionnaires could have been shortened and reviewed critically for absolutely necessary items. It is preferable to ensure an acceptable questionnaire fill-in time at the expense of not answering all posed research questions[782]. Also, some questions and topics are considered sensitive, so that a questioned person would prefer not to answer to strangers. Depending on different respondents, these topics can include the personal financial situation, judgments about other people, personal flaws, or wrong behavior. In those cases, answers are denied, over- or understated, evasive answers given or lies told[783]. In addition, people tend to present a favorable image of themselves on questionnaires; a phenomenon called socially desirable responding[784]. Therefore, sensitive questions should best be avoided or asked in an attenuated form. Participants should be assured that all responses are evaluated confidentially and anonymously and that no personal data will be revealed[785]. Another way of attenuation is to change direct into projective questions, which are formulated in a way that makes respondents believe that they are evaluating another person's behavior or actions. This facilitates honest answers to sensitive questions while the answers can actually be projected on the respondent's behavior[786].

---

[777] Cooper/Schindler (2014), p. 225; Hüfken (2019), pp. 760-763. For the risk of social desirability, cf. the corresponding part in section 6.2.2.

[778] Berekoven et al. (2009), pp. 94-98.

[779] Porst (2014), p. 16.

[780] Cooper/Schindler (2014), p. 304.

[781] Döring et al. (2015), p. 410.

[782] Döring et al. (2015), p. 410.

[783] Berekoven et al. (2009), p. 97.

[784] Van de Mortel (2008), pp. 40-41.

[785] Porst (2014), p. 131.

[786] Gröppel-Klein/Königstorfer (2009), p. 541.

Regarding *sequencing*, in principle composition of questionnaires does not follow a fixed scheme so that researchers can freely design and formulate questions as well as develop suitable questionnaire layouts[787]. However, certain sequencing principles can encourage participant commitment and foster rapport. Interesting questions early on awake interest and motivate participation, so that classification questions like age or sex should instead be asked at the end of a survey[788]. Especially the opening question determines whether a target person will respond to the survey as once the first few questions are answered, participants tend to feel bound to complete a questionnaire. Therefore, this question should be exciting, topic-related, personally affecting a respondent, and be answerable by any participant[789]. Moreover, questions should be arranged in modules that structure the questionnaire thematically, as shifting between various topics confuses respondents, causes misunderstandings, and leads to less reliable answers[790]. A good practice is to arrange questions from more general topics to more specific ones. This helps to avoid the risk of respondent's bias, for instance, if a specific behavior is asked first, followed by general behavioral questions, and allows a respondent to ease into a subject, i.e., recall their overall behavior before getting questioned about specifics[791]. The order in which questions are posed can influence survey participants' answers, which is referred to as "response order effects[792]." For questions where individual answers need to be selected, respondents tend to select the first choices shown; at oral interviews, participants select the last choices they hear[793]. To counteract such biases, software tools for computer-assisted surveying can be programmed in a way that each participant receives answer alternatives in randomized or reversed order[794]. Furthermore, *data types and scales* have to be considered. Questions can be raised as closed questions, where all answer alternatives are provided, and open questions, where answering options are not included. Closed questions are faster to raise and analyze but require that the spectrum of possible answer categories is known. Conversely, open questions are more costly and time-consuming to analyze but can lead to more comprehensive, differentiated outcomes[795]. Responses for closed questions can be measured using the four data types "nominal," "ordinal," "interval," and "ratio." These are also frequently described as types of measurement scales[796]. Nominal scales are used for data that can be classified, i.e., have mutually exclusive and collectively exhaustive categories, but have no order, distance, or natural origin[797]. They are a qualitative assignment of characteristics of a variable, such as marital status, that can be differentiated into single, married, and widowed, which can later be used to categorize respondents[798]. Ordinal scales are for data that can be classified and have an order, but no distance or natural origin. They are used when the lesser or greater value should be determined[799]. An example is the ranking at a contest where a smaller-larger relation, such as first place, second place, can be made[800]. Interval scales are used for data that can be

---

[787] Schäfer (2016), pp. 29-30.

[788] Cooper/Schindler (2014), pp. 317-320.

[789] Porst (2014), pp. 139-146.

[790] Fietz/Friedrichs (2019), pp. 814-816.

[791] Brace (2008), pp. 40-42.

[792] Fietz/Friedrichs (2019), pp. 817-818.

[793] Artingstall (1978), pp. 1125-1138; Porst (2014), pp. 137-139.

[794] Cooper/Schindler (2014), p. 310.

[795] Fietz/Friedrichs (2019), p. 818; Reinecke (2019), pp. 720-721. Sometimes "hybrid questions" are also distinguished. They also have a set of predefined answers, but additionally offer that an open answer is provided, for instance in the category "other," cf. Reinecke (2019), pp. 720-721.

[796] Brace (2008), pp. 59-65.

[797] Cooper/Schindler (2014), pp. 249-254.

[798] Blasius/Baur (2019), pp. 1382-1383.

[799] Cooper/Schindler (2014), pp. 249-254.

[800] Hollenberg (2016), p. 31.

classified, ordered, and have an equal distance, but no natural origin. They are used to determine equality of intervals or differences[801]. For example, the Celsius scale is an interval scale as the temperature difference between 2 °C and 4 °C is the same as between 23 °C and 25 °C, and the corresponding intervals reflect the same temperature difference[802]. Ratio scales are used for data that can be classified, ordered, have an equal distance, and a natural origin. They are used for the determination of equality of ratios[803]. Ratio scales have a natural zero, for instance, as with age or measures of weight[804]. The most commonly used scaling type is the Likert scale, also known as the "Agree-Disagree scale[805]." With this type, respondents are presented with a series of attitude dimensions, a question battery, for which they should indicate how much they agree or disagree[806]. Traditionally, five levels of agreement are distinguished, although Likert-type scales also use seven and nine scale points. The Likert scale can be easily and quickly constructed, is more reliable, and provides greater data volumes than other scale types[807]. Measurement theory considers answers on Likert scales as ordinally scaled, although in research practice, it is considered as quasi-metric and can therefore be analyzed like an interval scale[808]. Concerning the number of scale points, more scale points generally allow respondents to better discriminate among choices, while too many go beyond their capacity to discriminate and therefore result in measurement errors[809]. Studies showed that the number of response categories should be ideally seven plus/minus two, and odd. Furthermore, for scales with seven or more categories, answer categories should be individually labeled. Besides, the order must alternate between positive and negative categories[810].

These recommendations were incorporated into the creation of the survey instrument for this study. The wording was optimized, and terms precisely defined in a culturally neutral way. To further ensure that the study setting is understandable, an illustration of the personal relationship in focus was shown on the initial page. Hints were given that all responses were evaluated confidentially and anonymously and that no personal data would be revealed. In addition, questions were posed as projective questions. A seven-point scale, either as Likert or as a two-dimensional scale where needed, was employed. Questions were posed in topic blocks, and the sequence was following a logical flow. Also, the used survey tools allowed for a rotation of questions. The questionnaire's initial version was tested and subsequently reviewed by two supply chain professionals for clarity, familiarity of terms, and briefness. Based on their suggestions, the questionnaire was modified.

Survey conduction

The original version of the final questionnaire was drafted in English. To *translate* it, the following common approach was used[811]. Two native Chinese speakers with excellent command of the English language were asked to translate it into Chinese. The two Chinese speakers then

---

[801] Cooper/Schindler (2014), pp. 249-254. For a further discussion about types of interval scales, measuring theoretical challenges and applications cf. Döring et al. (2015), pp. 244-256.
[802] Döring et al. (2015), p. 244.
[803] Cooper/Schindler (2014), pp. 249-254.
[804] Hollenberg (2016), p. 31.
[805] Döring et al. (2015), p. 269.
[806] Brace (2008), pp. 73-76.
[807] Cooper/Schindler (2014), pp. 278-279.
[808] Völkl/Korb (2018), p. 20.
[809] Komorita/Graham (1965), pp. 987-994.
[810] Franzen (2019), p. 851.
[811] Bullinger (1995), 1360-1362.

discussed and agreed on a common Chinese version of the questionnaire. This was back-translated into English by an independent English speaker with excellent knowledge of Chinese. His translated version was subsequently reviewed by the Chinese speakers and discussed for conceptual equivalence and linguistic performance, i.e., does the translation involve common language. Problematic Chinese terms were revised, and the resulting version was pretested by a Chinese supply chain professional who also evaluated it for clarity. Based on her input, the Chinese and English speakers jointly developed a Chinese version of the questionnaire, which was optimized for conceptual equivalence, clarity, and linguistic performance. The same approach was also used to create a German version of the questionnaire. All questionnaires are shown in Appendix s1.

Regarding the *contact approach*, it must be considered that a questioned person has to have enough motivation to engage with the survey questions[812]. This degree of motivation depends on the importance that a respondent attaches to survey results as well as the required effort. If a minimum level of motivation is not achieved, a questionnaire will not be started, interrupted in between, or not truthfully answered[813]. Following recommendations to maximize online survey participation, every potential respondent was personally addressed with an invitation email that provided background information and highlighted the importance of the study[814]. All participants were assured that no individual answers or personal data would be revealed. For the English and German versions of the questionnaire, the widespread *survey tool* "QuestionPro" was used as it offers a user-friendly interface, helpful layout customization, and visual analysis tools[815]. Due to accessibility constraints of QuestionPro in China, the survey tool "Wenjuan Xing" (WJX) was adopted. It is easily reachable and popular in China and provides similar functionality as QuestionPro[816].

6.2.3. Participants

This section will describe the sampling approach and introduce the respondents' structure, classified according to personal and firm characteristics.

Sampling

Sampling, drawing samples from a target population, is an essential step in conducting surveys[817]. The advantages of sampling compared to a full investigation into the target population are lower costs, greater result accuracy, greater speed of data collection, and the availability of population elements[818]. In addition to defining the population, key population parameters, i.e., summary descriptors of variables of interest in the population, should be laid out[819]. A probability sampling approach is often suggested, where the selection rules ensure that interferences to the target population from which the sample was drawn can be estimated[820]. In that case, every participant has an equal probability of being selected from the

---

[812] Hollenberg (2016), p. 1.

[813] Hlawatsch/Krickl (2019), pp. 357-359.

[814] Wagner-Schelewsky/Hering (2019), p. 789.

[815] Rosa/Toledo (2012), pp. 141-142.

[816] Mei/Brown (2017), pp. 722-730.

[817] Wagner-Schelewsky/Hering (2019), pp. 790-793.

[818] Cooper/Schindler (2014), pp. 338-341. Although counterintuitive, sampling allows for "better interviewing (testing), more thorough investigation of missing, wrong, or suspicious information, better supervision, and better processing" than possible with a census, cf. Deming (1990), p. 26.

[819] Cooper/Schindler (2014), pp. 345-346.

[820] Tansey (2007), pp. 768-770. For an overview about sampling techniques, cf. Cooper/Schindler (2014), pp. 349-361.

population[821]. This can assure that even when a study is repeated on different samples selected from a given population, the findings would not differ beyond specific error ranges[822]. Challenges in probability sampling are that the total population may not be available for a study and that questionnaire receivers can decide whether they want to respond[823]. An alternative approach is nonprobability sampling, where samples from a larger population are drawn subjectively by the researcher[824]. It is frequently used in social science research, although technical challenges limit the extent to which valid inferences to a population can be made. It is impossible to estimate the probability an element has to be included in a sample and the extent of sampling errors[825]. Nonprobability sampling procedures are used because they might satisfactorily meet the sampling objectives or for cost and time concerns. For instance, there might be no intention to generalize to a population parameter if the intention is to uncover a range of conditions or dramatic variations[826]. Nonprobability sampling can be unrestricted, purposive, or conducted following the snowball technique[827]. In unrestricted samples, so-called samples of convenience, the most readily available respondents are selected regardless of characteristics. This is opportune but eliminates the possibility to identify what population the sample group represents and how it might differ from other samples[828]. In purposive sampling, the study's purpose and the researcher's knowledge of the population guide the sampling process. This assumes that with good judgment and an appropriate strategy, researchers can select particular respondents of interest that are deemed most appropriate and thus develop samples that suit their needs[829]. In snowball sampling designs, an initial set of relevant respondents is identified and requested to refer the researcher to other potential subjects who might also be relevant to the object of study. It is a particularly useful method when respondents are challenging to identify and to contact, such as with power elites or lawbreakers[830]. Various sampling techniques can also be combined in creative ways in order to answer research questions[831]. Such mixed methods sampling strategies should be logically derived from the research questions addressed by the study and consider underlying assumptions of certain sampling techniques. Overall, the sampling approach should be ethical, feasible, and efficient[832].

As the research questions of this study are related to the selection process of LSPs, with a focus on the automotive industry, the target population are members from logistics and purchasing departments in the automotive industry, typical buying center members, and members of the operations and sales departments of LSPs, typical selling center members, that are working with automotive industry clients[833]. In line with the research questions, the key population parameter is the expertise in LSP selection processes. Input from experts in various regions is required to contrast regional differences. Both the automotive and the logistics service industry have

---

[821] Fink (2003), pp. 9-11.
[822] Judd et al. (1991), p. 132.
[823] Cooper/Schindler (2014), pp. 358-359.
[824] Tansey (2007), pp. 768-770.
[825] Feild et al. (2006), pp. 566-568.
[826] Cooper/Schindler (2014), pp. 358-359.
[827] Tansey (2007), pp. 769-770.
[828] Tansey (2007), pp. 769-770.
[829] Judd et al. (1991), p. 132.
[830] Cooper/Schindler (2014), pp. 360-361.
[831] Teddlie/Yu (2007), p. 77.
[832] Teddlie/Yu (2007), pp. 96-98.
[833] Pfohl (2016), pp. 345-347; Anderson et al. (2011), pp. 9-15.

thousands of market players worldwide[834]. Considering this vast number of companies and their varying organizational setups, it is unrealistic to define the size of the target population as required for probability sampling methods. Besides, the research topic is rather sensitive, and respondents might be difficult to contact. Therefore, a mixed method sampling approach combining purposive sampling and snowball sampling was selected. First, four regional clusters of countries with a sufficiently big automotive industry and cultural and regional proximity were defined. These are "Western Societies," "Latin America," "China," which is considered an independent regional cluster due to its size and the research focus, and "Rest of Asia." Then, due to close cooperation with a leading automotive Tier 1 supplier, contacts to buying center members in every regional cluster were obtained. They were contacted and asked to provide references to their LSP counterparts who might be suitable to provide input for the survey. The survey was then initially distributed to 1,876 potential participants (English and German version: 1,432, Chinese version: 444), who then forwarded it to an unknown number of possible respondents.

Respondents

The survey was online available from 2018-06-03 to 2018-06-29, started 512, and completely filled in 194 times. This represents a drop-out rate of 62%, which might be related to the topic's sensitivity. It took participants, on average, 8.5 minutes to fill in the survey. What follows is a description of the respondents' structure according to their personal and firm characteristics.

<u>Personal Characteristics</u>

Regarding *sex*, 80 female and 106 male participants completed the survey. Eight participants did not disclose their sex. With regard to *regional clusters*, "China" included 94 respondents, "Western Society" 53, employed in Germany (25), the US (9), Romania (8), Czech Republic (5), Netherlands (3), Portugal (3), and Belgium (1), "Latin America" 28, working in Brazil (23), Venezuela (2), Chile (1), Colombia (1), and Mexico (1). The cluster "Rest of Asia" consisted of 16 respondents from India (6), Thailand (5), Singapore (3), and Japan (2). Three respondents did not provide information about their working location. Figure 8 displays the regional clusters of the respondents.

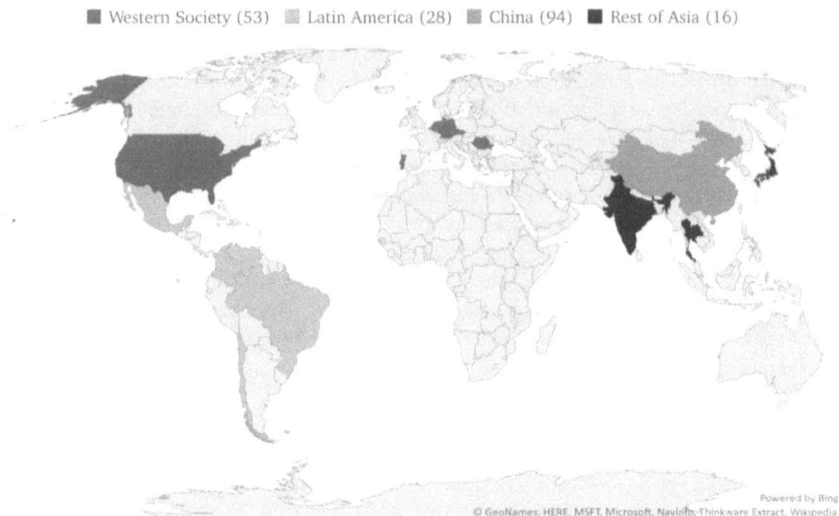

Figure 8: Survey respondents regional clusters.
Source: Own figure.

On average, the participants have 13 years of *work experience* in logistics and supply chain management. To allow for further data analysis, the sample was divided into three clusters of equal size, based on the 33rd and 66th percentiles of their years of work experience. The first cluster, "Little work experience," ranged from 1-8 years of work experience and included 64 participants (33%), the second cluster "Medium work experience" from 9-15 years with 66 participants (34%) and the third one "Much work experience" included respondents with above 16 years of work experience which was the case for 57 participants (29%). Seven (4%) participants did not provide information about their work experience. The results are summarized in Table 3. The frequency peaks at 5, 10, 15, 20, 25, and 30 years are likely an indication that some respondents rounded their years of work experience up or down. To confirm respondents' knowledge in the research topic, the participants were asked to evaluate their *expertise in the selection process of LSPs* on a scale from 1 (low expertise) to 7 (high expertise). Four participants rated their expertise as 1, "low expertise" (2%), 15 as 2 (8%), 30 as 3 (15%), 40 as 4 (21%), 51 as 5 (26%), 32 as 6 (16%) and 19 as 7 "high expertise" (10%). Three (2%) did not provide an answer. The mean value is 4.5, which allows for the conclusion that most respondents have sufficient proficiency in the research subject and can provide relevant responses. All personal characteristics are summarized in Table 3.

Table 3: Survey respondents personal characteristics

Characteristic	Distinction		Respondents
Sex		Female	80
		Male	106
		No answer	8
Region	China	84	94

	Western Society	Germany	25
		United States of America	9
		Romania	8
		Czech Republic	5
		Netherlands	3
		Portugal	2
		Belgium	1
	Latin America	Brazil	23
		Venezuela	2
		Chile	1
		Colombia	1
		Mexico	1
	Rest of Asia	India	6
		Thailand	5
		Singapore	3
		Japan	2
	No answer		3
Work experience [years]	**Little**	1	4
		2	4
		3	8
		4	8
		5	14
		6	7
		7	9
		8	10
	Medium	9	4
		10	17
		11	7
		12	10
		13	8
		14	8
		15	12
	Much	16	5
		17	3
		18	2
		19	4
		20	14
		21	3
		22	3
		23	3
		24	2
		25	4
		26	1
		27	1
		28	2
		30	7
		33	1
		35	1

	40	1
No answer		7
Expertise in LSP selection process	1 (low expertise)	4
	2	15
	3	30
	4	40
	5	51
	6	32
	7 (high expertise)	19
	no answer	3

<u>Firm Characteristics</u>

Regarding the *firm size*, with 175 respondents, a 90% majority is working for large enterprises. Fifteen respondents (8%) are working for SMEs, and 4 (2%) did not provide a response. This uneven sample distribution must be considered in the analysis, and results concerning the company size interpreted cautiously. Concerning the *internationalization degree*, 113 respondents (58%) are working in multinational enterprises, 77 (40%) in domestic companies, and 4 (2%) did not answer. Concerning the *firm type*, most respondents (154, 79%) are working for a buying company, while 24 (12%) work for an LSP. Three participants (2%) responded that they are employed in other industries, specifically the airline industry (1), the consumer goods industry (1), and the power tools (1) industry. This might be related to the complex business unit structure of the Tier 1 supplier that provided the initial contacts, or the snowballing. 13 respondents (7%) did not provide an answer regarding their firm type. This unequal respondent distribution needs to be considered when interpreting results regarding the firm type. Relating to the *selection process structuredness*, survey participants were asked to rate how structured they perceive their firm's process on a scale from 1, very unstructured, to 7, very structured. Out of the 191 respondents that provided an answer (3 respondents, which corresponds to 2%, did not), two rated it as 1 (1%), 10 as 2 (5%), 14 as 3 (7%), 28 as 4 (14%), 51 as 5 (26%), 46 as 6 (24%) and 40 as 7 (21%). The mean value is 5.17, indicating that most survey participants consider the process in their company rather structured. All firm characteristics are summarized in Table 4.

Table 4: Survey respondents firm characteristics

Characteristic	Distinction	Respondents
Size	Small and medium sized company	15
	Large enterprise	175
	No answer	4
Internationalization degree	Multinational enterprise	113
	Domestic company	77
	No answer	4
Firm type	Logistics Service Provider	24
	Buying company	154
	Other	3
	No answer	13

Selection process	1 (very unstructured)	2
	2	10
	3	14
	4	28
	5	51
	6	46
	7 (very structured)	40
	No answer	3

6.2.4. Analysis

The next section outlines the analysis techniques used in this study, introduced by a section about given constraints regarding statistical hypothesis testing feasibility.

Preceding considerations

To verify theoretical assumptions, that is to accept or refute a null hypothesis, statistical hypothesis testing, also called confirmatory data analysis, is mandatory[835]. In the current study, a nonprobability sampling approach instead of probability sampling was used. As samples were not randomly selected, a normal distribution of the sample cannot be assumed. This is required for a confirmatory data analysis, which is therefore not feasible[836]. Also, some concepts of the qualitative study are abstract. To operationalize them for hypothesis testing, they would have to be translated into various concrete, measurable variables. As those variables would have to be queried individually, it would result in an impractically high number of questionnaire items. Some are also not clearly defined and would require the extensive development of operational definitions to measure such constructs' characteristics[837]. This would go beyond the aim of this study and stretch the motivation of the participants unduly. As a result, all relevant variables were directly included in the questionnaire. Although this approach does not allow to prove hypotheses unambiguously, research propositions can nevertheless be verified. The necessary analysis was conducted in several steps. First, a descriptive analysis with frequencies, crosstabs, mean values, and a Mann-Whitney-U-Test about location differences was completed. Second, the interrelationships of the enquired items were examined using explorative factor analysis and regression analysis. To manage the data and carry out the statistical tests, the IBM SPSS Statistic software in version 25 was used[838].

Descriptive Analyses

Descriptive statistics combine all methods with which empirical data can be summarized and described[839]. Asides from frequencies and mean values, which are assumed to be well known, the descriptive methods for the analysis will be briefly explained[840].

---

[835] Homburg (2017), pp. 299-307; Tansey (2007), pp. 768-770.

[836] Tansey (2007), pp. 768-770.

[837] Homburg (2017), pp. 398-405; Yen et al. (2011), pp. 99-106.

[838] Wagner (2019), pp. 1-2.

[839] Schäfer (2016), p. 47.

[840] Schäfer (2016), pp. 47-50.

<u>Crosstabs</u>

Crosstabs also called contingency tables, show different combinations of values of nominally scaled variables. They allow examining relationships between two or more variables[841]. For example, it can be assessed whether the average value for the result of a particular group, such as male participants, is higher than that of another one, such as female participants.

<u>Mann-Whitney-U-Test</u>

The Mann-Whitney-U-Test (MWU-Test) can be used to statistically prove whether the results of two independent random samples actually differ from each other[842]. It is a non-parametric test that can be applied if the prerequisites for a parametric procedure, such as the t-test for independent samples, are not fulfilled[843]. For instance, it can also be calculated for small samples or outliers. Prerequisites are that the dependent variable is at least ordinally scaled, although it can also be used with interval and ratio measures, and that at least two comparable groups can be formed through an independent variable[844]. The MWU-Test compares at least two independent samples and determines their mean ranks[845]. If a systematic difference between the two groups exists, most low ranks will belong to one group while most high ranks will belong to the other one, and therefore the total mean ranks differ. The distance between the two mean ranks is described by the test variable "U." For large samples, it could be demonstrated that the U-value is approximately normally distributed. Therefore, it can be converted into a z-value and tested for significance[846]. The null hypothesis assumes that both tested groups originate from the same basic population. If the two mean ranks differ at a certain significance level, typically $p < 0.05$, it can be assumed that the two tested groups differ significantly from each other, and the null hypothesis must be rejected[847].

Explorative Factor Analysis

The Explorative Factor Analysis (EFA) is a multivariate data analysis method applied to discover unknown correlative structures in data sets[848]. EFA attributes a large number of observed, or manifest, variables to a few underlying or latent factors[849]. Through this technique, data are condensed and can subsequently be easier interpreted, uncovering underlying causes[850]. The EFA analyzes relationships in large sets of variables by identifying groups of highly correlated variables and separating them from less correlated groups[851]. The groups of highly correlated variables form the latent variables, so-called factors[852]. As the target of the EFA is to discover underlying factors, different from the confirmatory factor analysis, which aims at confirming predefined factors and assigned variables, no factors are defined beforehand and only specified after the analysis was completed[853]. The EFA is conducted in five steps[854]. First, the *creation of*

---

[841] Schäfer (2016), pp. 249-250.
[842] Nachar (2008), p. 15.
[843] Schäfer (2016), p. 244.
[844] Cooper/Schindler (2014), pp. 615-617.
[845] Schäfer (2016), p. 243.
[846] Cooper/Schindler (2014), pp. 615-617.
[847] Cooper/Schindler (2014), pp. 615-617.
[848] Backhaus et al. (2016), p. 385.
[849] Kuß et al. (2018), pp. 295-298.
[850] Magerhans (2016), pp. 157-160.
[851] Magerhans (2016), pp. 157-160.
[852] Backhaus et al. (2016), pp. 385-386.
[853] Magerhans (2016), pp. 157-160.
[854] Homburg (2017), pp. 364-371; Kuß et al. (2018), pp. 268-271.

a data matrix. A data matrix of objects and variables is to be created based on a sample with a sufficient number of variables. Data should be standardized to enable a comparison of variables from different scales[855]. Second, *calculation of a correlation matrix*. A correlation matrix should be created, which reveals similarities across variables[856]. It provides an initial overview of the data suitability, as it is questionable if the EFA will be feasible if the data structure is too heterogeneous, indicated by low correlations[857]. In this step, also, two statistical tests can be carried out to evaluate data suitability. "Bartlett's-Test of Sphericity" tests the hypothesis that a sample consists of a population in which variables are uncorrelated. It can be a lower bound for the process so that if the independence hypothesis cannot be rejected, the EFA does not need to be further pursued[858]. The "Kaiser-Meyer-Olkin (KMO)" criterion, also known as "Measure of Sampling Adequacy (MSA)," is considered as a standard test procedure to determine if data are suitable for the EFA. It indicates to which extent variables belong together, where a correlation matrix with MSA < 0.5 is considered unacceptable and MSA ≥ 0.8 as desirable for an EFA[859]. Third, *determination of number of factors*. Although there are no universal rules for determining the number of factors, various statistical tests, especially the "Kaiser criterion" and the "Scree test," are commonly used[860]. They are both based on the correlation matrix's eigenvalue, which indicates how much of the total variance of all variables is explained by one particular factor. The higher the eigenvalue of a factor, the greater the total variance explained by it. According to the Kaiser criterion, factors with an eigenvalue greater than 1.0 should be extracted. The scree plot is a graphical representation of the eigenvalue, where the number of factors is plotted on the x-axis, and the eigenvalues of the factors are plotted on the y-axis. Typically, it shows a line that first drops steeply and then flattens out, resembling a bent arm. At the point where the difference of the eigenvalues between two factors is greatest, like an elbow, the bend is visible. According to the corresponding "elbow criterion," the first point left of this bend determines the number of factors that are to be extracted. As both the Kaiser criterion and the Scree test do not necessarily lead to the same result, the ultimate decision about the number of factors must be based on the interpretability of the content[861]. Forth, *rotation and interpretation of the factors*. The interpretation is based on each variable's factor loadings, which indicate the strength of the correlation between the individual variables and the extracted factors. The factor loadings are presented in a factor loading matrix[862]. This matrix should be transformed into a "simple structure," where variables always load high on only one factor and low on all the other factors. This transformation is achieved through rotation techniques such as the orthogonal "Varimax rotation" method[863]. The subsequent interpretation itself is a subjective, theoretical, and inductive process, where the researcher must distill factors and assign them names or themes based on a number of high loading variables[864]. Here, factor loadings of 0.3 or higher, which explain approximately 10% of a variable's variance, should be considered[865]. Fifth, the *construction of factor scores*. Factor scores are linear combinations of all of the measures,

---

[855] Homburg (2017), pp. 364-371; Backhaus et al. (2016), pp. 385-386. In this study, questions Q2, Q5 and Q9 provide the input data for the EFA which results in a total of 17 tested variables (8 for Q2, 8 for Q5, and 1 for Q9). The variables are uniformly scaled on a seven-point Likert scale which can be analyzed like an interval scale Völkl/Korb (2018), p. 20.

[856] Homburg (2017), pp. 364-371.

[857] Backhaus et al. (2016), pp. 385-386.

[858] Dziuban/Shirkey (1974), pp. 358-361.

[859] Backhaus et al. (2016), 398-340.

[860] Homburg (2017), pp. 365-366.

[861] Kuß et al. (2018), pp. 295-298; Backhaus et al. (2016), pp. 405-417.

[862] Homburg (2017), pp. 365-366.

[863] Thurstone (1931), p. 407; Backhaus et al. (2016), pp. 417-420.

[864] Henson/Roberts (2006), p. 396; Williams et al. (2014), p. 9.

[865] Tinsley/Tinsley (1987), p. 422.

weighted by the corresponding factor loading[866]. Constructing them allows that factors can be used for further analysis, just like other variables in the data matrix[867]. This was also required for the next analysis step, a regression analysis.

Regression analysis with EFA-factor-scores

To examine the influence of independent variables on dependent variables, regression analysis (RA) is one of the most established methods[868]. In regression analyses, also called bivariate regression, statements about the strength and the significance of the influence of an independent variable (X) on a dependent variable (Y) can be made[869]. This requires that all used variables must be scaled metric or quasi-metric[870]. Variables where this premise is not given, e.g., demographic variables typically binary such as female/male, have to be converted into dummy variables (for example, female = 1, male = 0) that can then be treated as metric variables throughout the analysis[871]. EFA-Factor scores that were constructed as described in the previous section can be included as a dependent variable in a regression analysis. In this research, it led to a total of 12 independent variables summarized in Table 5. A separate simple regression calculation is performed for each possible combination of independent and dependent variables.

Table 5: Survey regression analysis variables

Question	Independent Variable X_i	Coding	Dependent Variable Y_i
Q8	Structured Selection Process	Likert	EFA-Factor scores of Factor 1
Q10	Firm Size small	Dummy	EFA-Factor scores of Factor 2
Q10	Firm Size large	Dummy	...
Q11	Multinational Enterprise	Dummy	
Q11	Domestic Company	Dummy	
Q12	LSP	Dummy	
Q12	Automotive Industry	Dummy	
Q13	Female	Dummy	
Q13	Male	Dummy	
Q14	China	Dummy	
Q14	Western Society	Dummy	
Q16	Years of work experience	Metric	

To measure how well the dependent variable can be explained by the independent variable ("goodness of fit"), the *coefficient of determination* (R-Square or R^2) is frequently used[872]. It is interpreted as the total proportion of variance in Y explained by X and shows how well the regression line fits the data[873]. The coefficient of determination is a normalized value ranging from zero, which means that none of the response variable variation is explained by the linear model, to one, i.e., the whole response variable variation is explained by the linear model.

---

[866] DeCoster (1998), p. 3.
[867] Kuß et al. (2018), p. 271.
[868] Cooper/Schindler (2014), pp. 479-480; Kuß et al. (2018), p. 279.
[869] Magerhans (2016), p. 141.
[870] Magerhans (2016), p. 141.
[871] Kuß et al. (2018), pp. 291-292.
[872] Stoetzer (2017), pp. 40-43.
[873] Cooper/Schindler (2014), pp. 489-490.

Values above 0.8 are considered desirable[874]. However, when interpreting this coefficient, it must be borne in mind that the value of R^2 is strongly dependent on the underlying problem and the type of data. For instance, macroeconomic time series data typically have a very high R^2 value above 0.9, as independent and dependent variables often have a common trend. In contrast, microeconomic problems that address individual people or companies' behavior often have a low R^2 value below 0.3, although the results are reasonably interpretable[875]. Also, in the current study, the size of R^2 is of less relevance, as regression analysis aims not to explain a maximum of the total variation, but instead to identify significant differences between various groups.

In the case of simple regressions with one independent and one dependent variable, the entire model's *significance value* corresponds to the significance value of the independent variable[876]. The significance values of the regression coefficients can be determined with the t-test[877]. The null hypothesis of the t-test states that the determined coefficient equals zero, which means that it actually does not affect the population[878]. The t-value is calculated by dividing the regression coefficient by their standard error. Through this value, the significance level (p-value) can be derived. Which significance level is deemed as acceptable should be defined before the analysis. Often, a value of 0.05 is used as a significance limit for the t-test and will, therefore, also be used in this study[879]. Hence, if $p < 0.05$, the null hypothesis must be rejected, and it can be assumed that the independent variable that is analyzed has a statistically significant influence on the dependent variable[880].

6.3. Results

The study results are presented in the following section. Subsequently, the outcome from Frequencies, Crosstabs, Mann-Whitney-U-Test, Explorative Factor Analysis, and Regression Analysis are shown.

6.3.1. Frequencies

This section will introduce the answer frequencies of the respondents. They are structured following the two main research questions RQs1 and RQs2.

RQs1: Which contingency factors influence the impact of personal relationships on the LSP selection?

The survey questions Q1, Q2, Q3, Q4 and Q9 are aimed at addressing RQs1, which contingency factors influence the impact of personal relationships on the LSP selection.

<u>Q1: Do personal relationships have any influence on the selection of a Logistics Service Provider?</u>

---

[874] Cooper/Schindler (2014), pp. 489-490; Backhaus et al. (2016), pp. 84-85.
[875] Stoetzer (2017), pp. 40-43.
[876] Backhaus et al. (2016), pp. 95-96.
[877] Cooper/Schindler (2014), pp. 487-488.
[878] Stoetzer (2017), pp. 45-46.
[879] Stoetzer (2017), pp. 45-46.
[880] Stoetzer (2017), pp. 45-46.

In the introduction question, participants should generally assess whether personal relationships influence the selection of LSPs. The majority of 129 respondents (69%) believe that personal relationships have an influence; 58 respondents (31%) believe that personal relationships do not have any influence. Seven did not respond to the question. Responses among the four "Country of work" regions vary considerably. In the cluster *China*, the share of people who assume that personal relationships influence the selection process of LSPs is at 82% higher than in the other clusters. At 62%, some 20 percentage points less, people working in *Western Societies* assume that personal relationships do have an influence. In *Latin America*, 46% believe that there is an influence, and in the *Rest of Asia,* 57%. Also, when comparing the various firm characteristics, differences are visible. Most of the employees of *small and medium sized companies* (85.7%), of *domestic companies* (85.7%), and of *Logistics Service Providers* (95.5%) assume that personal relationships have an influence on selection processes for LSPs. The opposing factors *large enterprise* (68.4%), *multinational enterprise* (57.4%), and *buying company* (65.6%), have significantly lower percentages. Likewise, the results divided based on personal characteristics vary. Regarding *sex*, no significant difference is observable. Only a marginally larger proportion of female participants (70.9%) than male participants (68.8%) assume that an influence of personal relationships exists. For the *years of work experience* characteristic, a clear tendency is visible. The lower the work experience, the higher the share of those who agree to the influence of personal relationships. In the *little work experience* group, 75.4% of the respondents agree, in the *medium work experience* group, 70.8%, and in the *lot of work experience* group, 58.9%. All results are summarized in Figure 9.

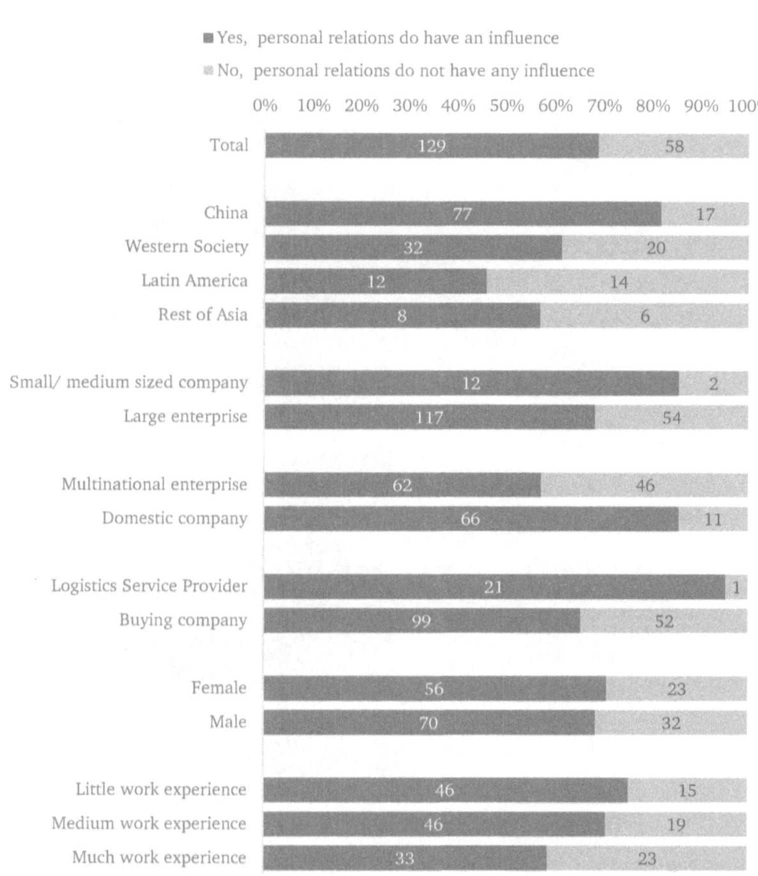

Figure 9: Have personal relationships any influence on LSP selection
Source: Own figure

Q2 How much do the following factors influence the extent of impact of personal relationships on the selection of Logistics Service Providers?

The second survey question addresses characteristics that might influence the decision-making process. Survey participants were asked to evaluate the different factors on a seven-point Likert scale from 1 (very weak influence) to 7 (very strong influence). Respondents rated all factors except of *H7 - Age of involved decision makers* as having a rather strong influence on the support possibilities. The most frequently selected option for H1 and H4 is "5," for H2 and H3, each "5" and "6" are equally chosen, and for H5, H6, and H8, the dominant answer is "6". All results are shown in Figure 10.

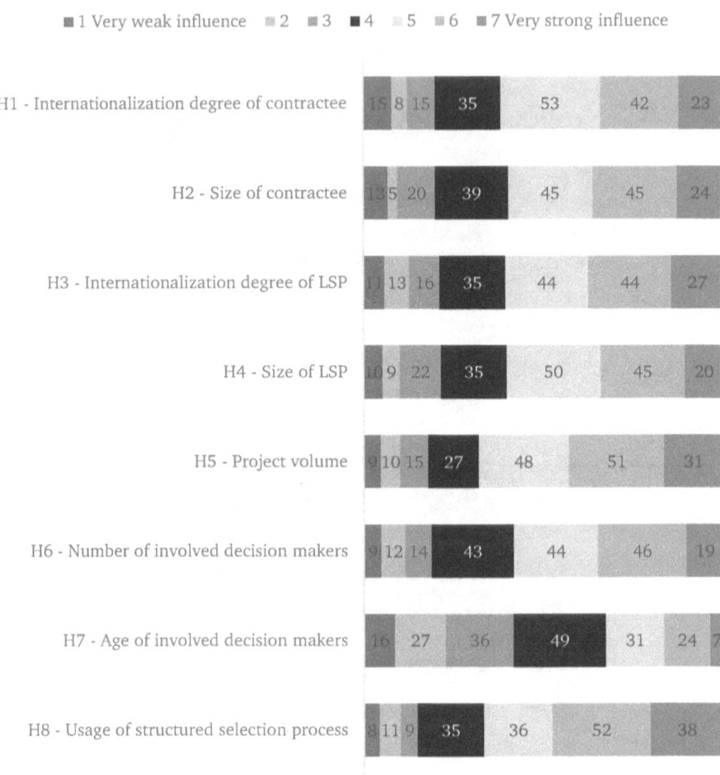

Figure 10: Influencing factors on impact of personal relationships on LSP selection
Source: Own figure

The differences among factors are easier to identify when comparing mean values as in Figure 11. Factors with the highest values are *H8 – Usage of structured selection process* with a mean of 5.05 and *H5 – project volume* with 4.95. Five factors are on a medium level. *H3 – Internationalization degree of LSP* (4.73), *H2 – Size of contractee* (4.72), *H1 – Internationalization degree of contractee* (4.68), *H4 – Size of LSP* (4.68), and *H6 – Number of involved decision makers* (4.68). The factor *H7 – Age of involved decision makers* is lowest with a value of 3.80.

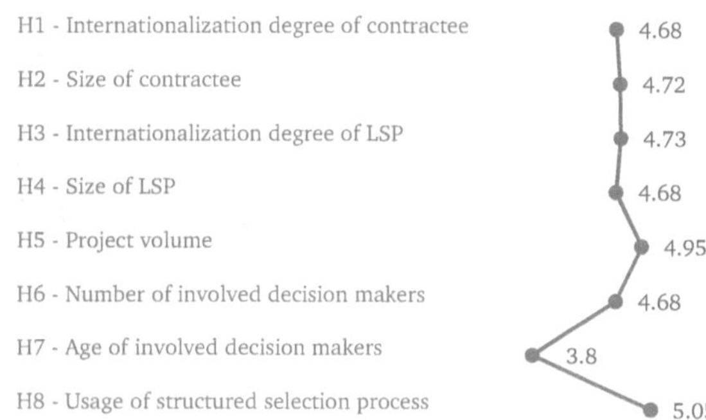

Figure 11: Mean values for influencing factors on impact of personal relationships on LSP selection
Source: Own figure

<u>Q4 Please select up to three factors that have the highest impact on how strong the influence of personal relationships on the selection of LSPs is</u>

In this question, survey participants should choose up to three factors that have the highest impact on how strong the influence of personal relationships on the selection of LSPs is. Consistent with the results of Q2, the factors *Project volume* and *Usage of structured selection process* reached the highest values with 100 mentions each, followed by the *Number of involved decision makers* with 82. With 27 mentions, the factor *Age of involved decision makers* was selected most seldom. The results are shown in Figure 12.

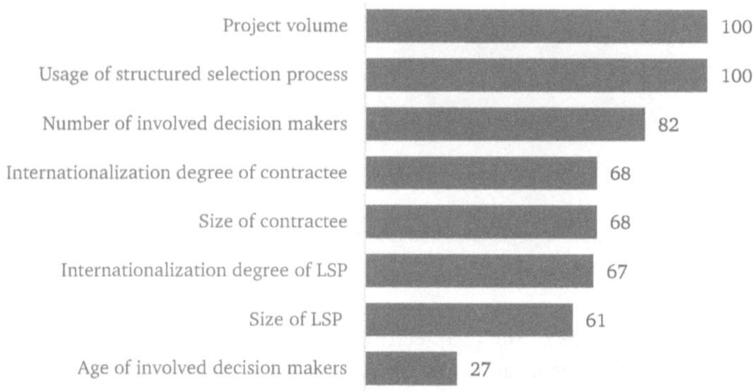

Figure 12: Highest influencing factors on impact of personal relationships on LSP selection
Source: Own figure

<u>Q3 Where do you think is the influence of personal relationships on the selection of LSPs greater? Please rate the opposing factors</u>

Q3 addresses the different opposing tendencies of the eight characteristics. A **seven-point** bipolar Likert scale is used with the extreme poles -3 and +3, and a center point of 0, allowing participants to take a neutral standpoint. The results are shown in Figure 13 and have to be interpreted as follows. For example, at the first factor *Internationalization degree of contractee* 36 respondents trust that if the contractee is a domestic company, the influence of personal relationships is much greater than if the contractee is a multinational enterprise (rating of -3). Eighteen people believe the opposite (rating of +3), and 32 people gave a neutral response (0). Overall, 86 participants, 36 + 31 + 19, assume that the influence of personal relationships is greater if the contractee is a domestic company, and 76 participants, 23 + 35 + 18, that it is greater if the contractee is a multinational enterprise.

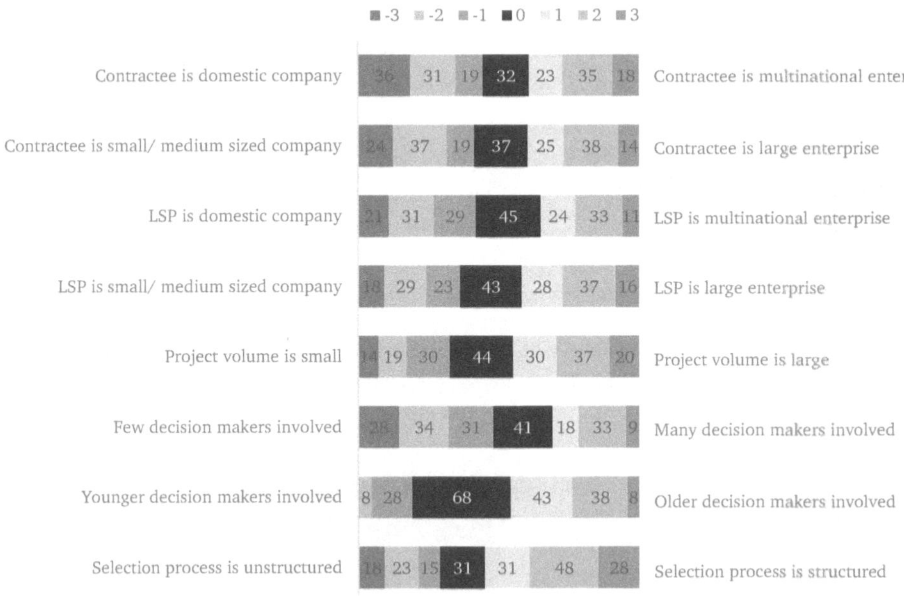

Figure 13: Opposing factors influencing impact of personal relationships on LSP selection
Source: Own figure

The mean values make the tendencies of the characteristics more apparent. For instance, for factor 5 *Project volume,* the mean value of all responses is 0.28, which translates into a larger share of respondents assuming that a larger project volume leads to a greater influence of personal relationships. All mean values are shown in Figure 14.

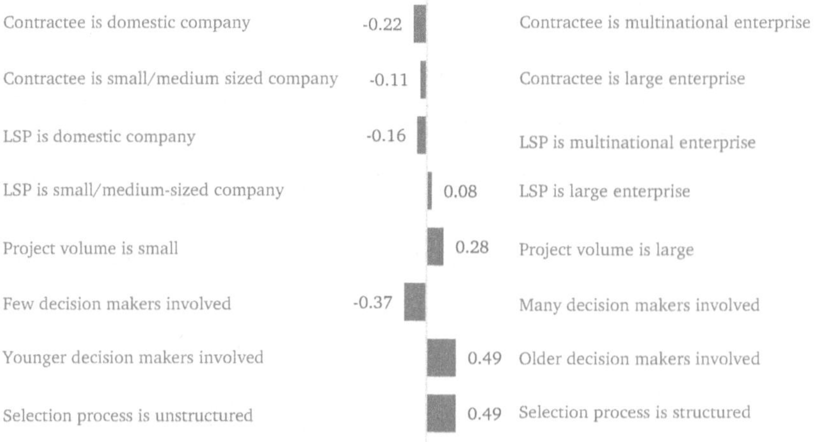

Figure 14: Mean values for opposing factors influencing impact of personal relationships on LSP selection
Source: Own figure

Q9 How strong is the influence of personal relationships on the selection of LSPs in your opinion?

Respondents should assess the strength of personal relationships' influence on the LSP selection on a seven-point Likert scale, ranging from 1 (very weak) to 7 (very strong). For example, ten people selected a very strong influence. The average value is 4.23, indicating an overall medium to strong influence. The results are shown in Figure 15.

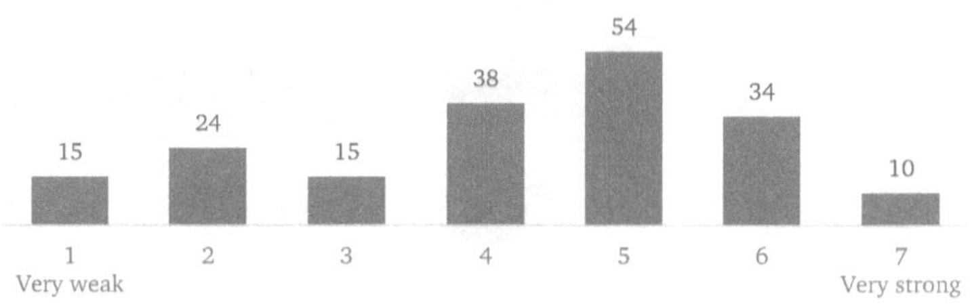

Figure 15: Strength of influence of personal relationships on LSP selection
Source: Own figure

RQs2: How do members of a Buying center support a certain LSP during the selection process?

The survey questions Q5 and Q6 address RQs2 - how do members of a buying center support a certain LSP during the selection process.

<u>Q5 Please rate how often the following situations occur because of personal relationships between associates from the Buying center of the contractee (e.g. logistics associate of Tier 1) and from the selling center of the Logistics Service Provider (e.g. key account manager)</u>

In Q5, survey participants should evaluate on a seven-point Likert scale from 1 (never) to 7 (very often), how often different support possibilities occur. For example, for the first support possibility *Buying center shares actual contract prolongation price reduction target*, 11 respondents believe that it occurs very often, 39 believe that it occurs with medium frequency, and 19 assume that it never occurs. Figure 16 provides an overview of all responses.

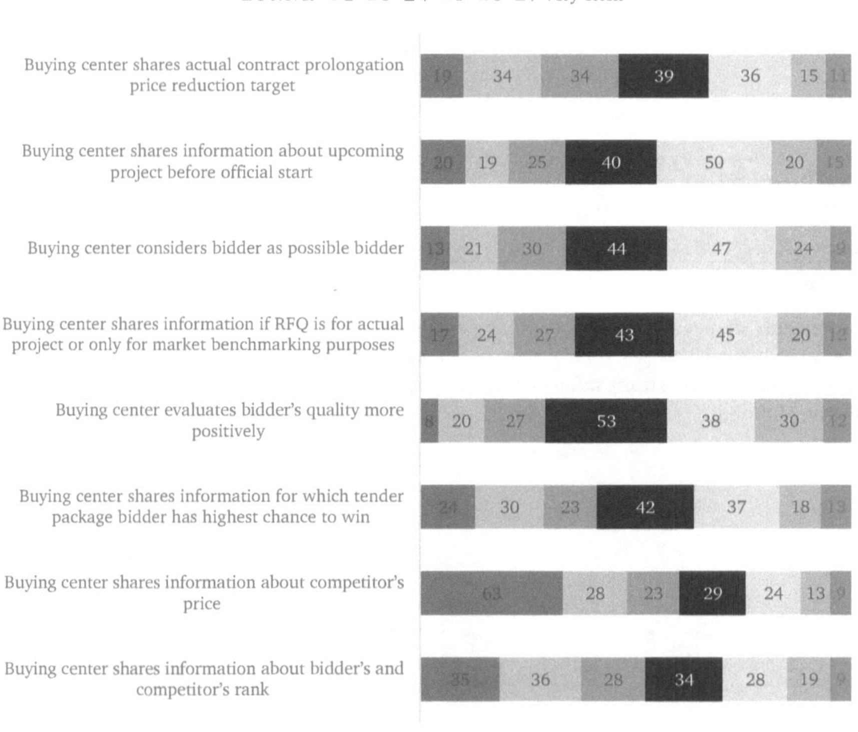

Figure 16: Occurrence of situations during LSP selection process due to personal relationships
Source: Own figure

The differences between the values become clearer through an analysis of the mean values. According to the study participants, the support possibilities *Buying center shares information about competitor's price* and *Buying center shares information about bidder's and competitor's rank* rarely occur, indicated by mean values of 2.99 and 3.41. Most commonly, a *Buying center shares information about upcoming project before official start* (4.06); a *Buying center considers bidder as possible bidder* (4.06), and *Buying center evaluates bidders quality more positively* (4.23). The results are summarized in Figure 17.

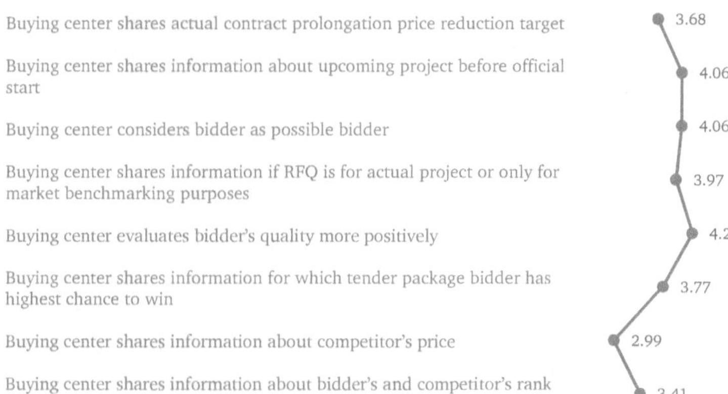

Figure 17: Mean values for occurrence of situations during LSP selection process due to personal relationships
Source: Own figure

Q6 Please select up to 3 situations which are most common in practice

Participants should select up to three situations which occur most commonly in practice. Most common are *Buying center shares information about upcoming project* (93 mentions), *Buying center evaluates bidders quality more positively* (79) and *Buying center considers bidder as possible bidder* (74). Least common are *information sharing about competitor's price* (31) and *information sharing about bidder's and competitor's rank* (42). This largely confirms the outcome of Q5. All results are shown in Figure 18.

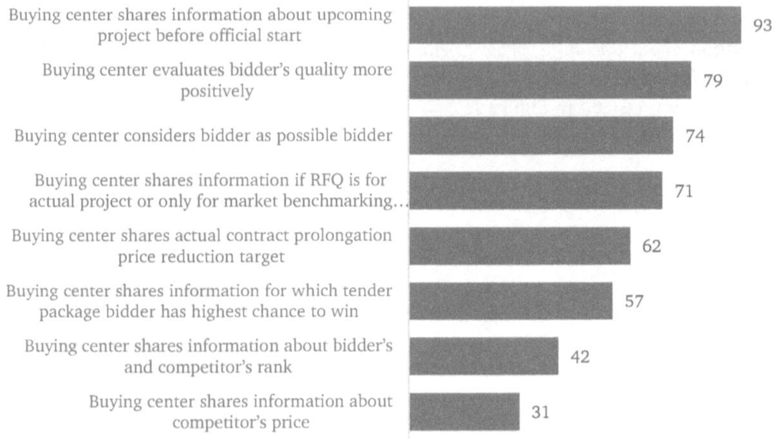

Figure 18: Most common situations during LSP selection process due to personal relationships
Source: Own figure

6.3.2. Crosstabs

In this section, various Crosstab analyses will be presented. After a total overview, the crosstabs China compared to Western societies, SMEs compared to large enterprises, Multinational enterprises compared to Domestic companies, LSPs compared to Buying companies, Females compared to Males, and Little compared to Medium compared to Much work experience will be examined.

Total

To identify differences of the influence of personal relationships on the LSP selection process between groups, mean values of the related questions Q2, Q5, and Q9 are contrasted. These questions were measured on a seven-point Likert scale with mean values ranging from 2.99 to 5.05. The results are shown in Table 6.

Table 6: Differences of personal relationship influence among groups

Group	Q2	Q5	Q9	Total	Difference
China	5.05	4.06	4.61	4.55	Δ 0.71
Western Societies	4.22	3.43	4.02	3.84	
SMEs	4.43	4.04	4.67	4.26	Δ 0.04
Large Enterprises	4.68	3.75	4.21	4.22	
Multinational Companies	4.35	3.55	3.88	3.95	
Domestic Companies	5.12	4.08	4.74	4.61	Δ 0.66

LSPs	4.69	4.42	5.12	4.58	Δ 0.39
Buying companies	4.68	3.71	4.10	4.19	
Females	4.70	3.70	4.20	4.20	
Males	4.65	3.85	4.24	4.25	Δ 0.05
Little work experience	4.81	3.83	4.48	4.33	Δ 0.13
Medium work experience	4.68	3.73	4.20	4.20	Δ 0.09
Much work experience	4.51	3.73	3.96	4.11	
Total	4.66	3.77	4.23	4.22	

Comparing the individual questions, the mean value of all eight factors queried in Q2 is 4.66; the mean value for the frequency of occurrence of the eight support possibilities queried in Q5 is 3.77, and the mean value of Q9 is 4.23. The factor with the highest rating across these questions is *Usage of structured selection process*; the factor with the lowest *Buying center shares information about competitor's price*. The values for all related figures are shown in Figure 19.

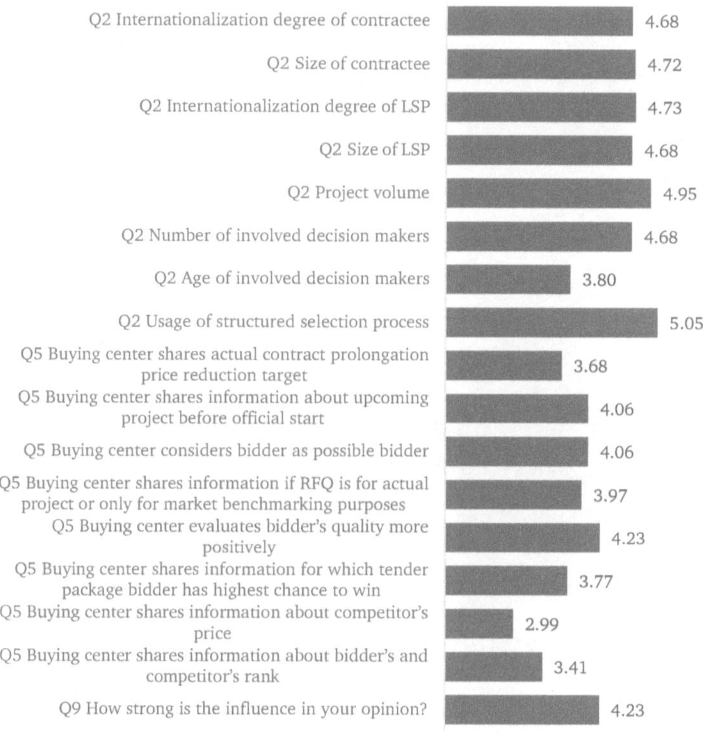

Figure 19: Means of personal relationship influence questions (Q2, Q5, Q9) across all groups
Source: Own figure

China compared to Western Societies

When comparing China and Western Societies, the mean values in China are consistently higher. For instance, in Q2, the mean value of the cluster China is 5.05, while in Western Societies, it is 4.22. A similar difference can be observed for Q5, where China's values are significantly higher (4.06) compared to Western Societies (3.43). Across all related questions, the average evaluation of people working in China is 0.71 points higher than the average evaluation from people working in Western Societies. The results are shown in Table 7.

Table 7: Differences of personal relationship influence by focus regions

Group	Q2	Q5	Q9	Total	
China	5.05	4.06	4.61	4.55	Δ 0.71
Western Societies	4.22	3.43	4.02	3.84	
Total	4.66	3.77	4.23	4.22	

This is also visible when comparing the results on a per-question level. Mean values for China are notably higher for Q2 *Internationalization degree of contractee* (Δ 1.13), and at Q5 for the support possibilities *Buying center shares information about competitor's price* (Δ 1.2), and *Buying center shares information about bidder's and competitor's rank* (Δ1.12). The results are shown in Figure 20.

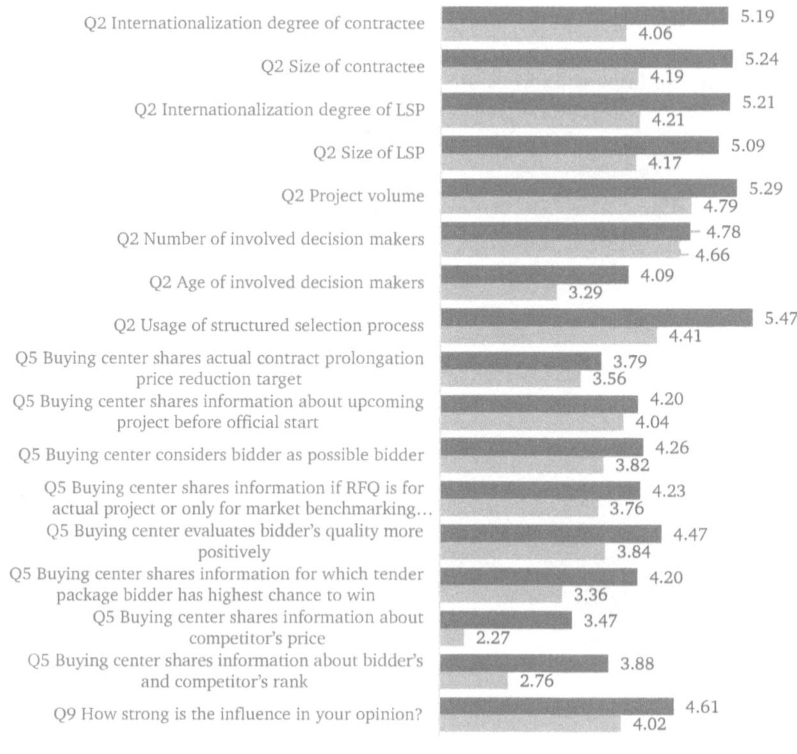

Figure 20: Means of personal relationship influence questions (Q2, Q5, Q9) for China compared to Western Society
Source: Own figure

SMEs compared to Large enterprises

Contrasting results for different firm sizes shows that SMEs' overall mean value is at 4.26 slightly higher than the one for large enterprises with 4.22. In Q2, mean values for SMEs are 4.43 and for large enterprises 4.68, while in Q5, the mean value of SMEs is with 4.04 slightly higher than the average of large enterprises with 3.75. The results for all related questions are shown in Table 8.

Table 8: Differences of personal relationship influence by firm size

Group	Q2	Q5	Q9	Total	
Small and medium sized Companies	4.43	4.04	4.67	4.26	Δ 0.04
Large Enterprises	4.68	3.75	4.21	4.22	
Total	4.66	3.77	4.23	4.22	

Figure 21 shows the differences between the firm sizes on a factor level. The highest mean values for SMEs are with the answers *Project volume* (5.2) in Q2 and *Buying center shares information for which tender package bidder has highest chance to win* (4.92) in Q5. The highest mean values for large enterprises are *Usage of structured selection process* (5.08) in Q2, and *Buying center evaluates bidder's quality more positively* (4.23) in Q5.

Figure 21: Means of personal relationship influence questions (Q2, Q5, Q9) for SMEs compared to Large enterprises
Source: Own figure

Multinational enterprises compared to Domestic companies

Comparing the results of respondents from multinational enterprises and domestic companies, a clear trend is observable. In all 17 surveyed situations, the mean value of people working in domestic companies is higher than that of multinational enterprises. This difference is of similar magnitude as that for China versus Western societies. The average value of multinational enterprises (4.35) for Q2 is 0.77 points lower than that of domestic companies (5.12). In Q5, multinational enterprises rated, on average, 3.55 and domestic ones 4.08. This is similar to Q9 with values of 3.88 and 4.74, respectively. The total difference across all questions is 0.66. All results are shown in Table 9.

Table 9: Differences of personal relationship influence by internationalization degree

Group	Q2	Q5	Q9	Total	
Multinational Companies	4.35	3.55	3.88	3.95	
Domestic Companies	5.12	4.08	4.74	4.61	Δ 0.66
Total	4.66	3.77	4.23	4.22	

Correspondingly, the breakdown across individual questions, presented in Figure 22, shows that the mean values for employees from domestic companies are consistently greater than those of multinationals. The most significant difference between domestic and multinational companies is for Q2 at the factor *Internationalization degree of LSP* (Δ1.14) and for Q5 at the support possibility *Buying center shares information about competitor's price* (Δ0.88).

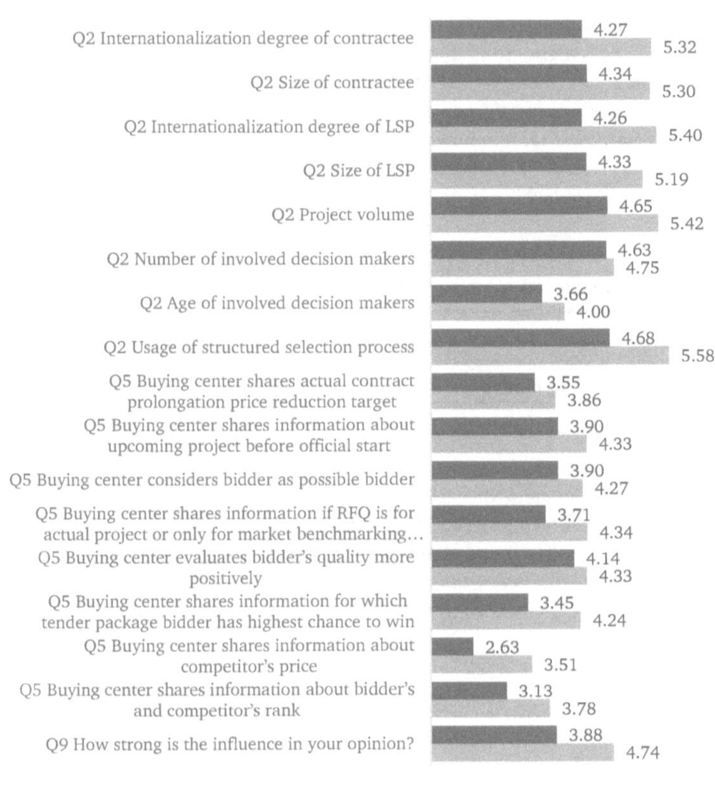

Figure 22: Means of personal relationship influence questions (Q2, Q5, Q9) for Multinational enterprises compared to Domestic companies
Source: Own figure

LSPs compared to Buying companies

Comparing Logistics Service Providers and Buying companies results, results differ between Q2, Q5, and Q9. In Q2, values are almost identical with a mean value of 4.69 for LSPs, and 4.68 for Buying Companies. However, for the actual support possibilities during the selection process (Q5), the LSPs' mean value is much higher (4.42) compared to Buying companies (3.71). Likewise, in Q9, the LSPs provided a rating of 5.12, while Buying companies 4.10, which is a difference of 1.02. Across all related questions, the mean value for LSPs is 0.39 higher than that of Buying companies (Table 10).

Table 10: Differences of personal relationship influence by firm type

Group	Q2	Q5	Q9	Total	
LSPs	4.69	4.42	5.12	4.58	Δ 0.39
Buying companies	4.68	3.71	4.10	4.19	
Total	4.66	3.77	4.23	4.22	

On a per factor level, these differences become more pronounced. LSPs rate the support possibilities *Buying center shares actual contract prolongation price reduction target,* and *Buying center shares information about upcoming project before official start* much higher than the Buying companies, with a difference of 1.16 and 1.24, respectively. Also, the LSPs' value for Q9 (5.12), *How strong is the influence in your opinion?*, varies noticeably from the Buying companies (4.10). The highest value for Buying firms is reached at Q2 *Usage of structured selection process* with 5.16. This factor is ranked 0.41 points lower by LSPs at 4.75. All results are shown in Figure 23.

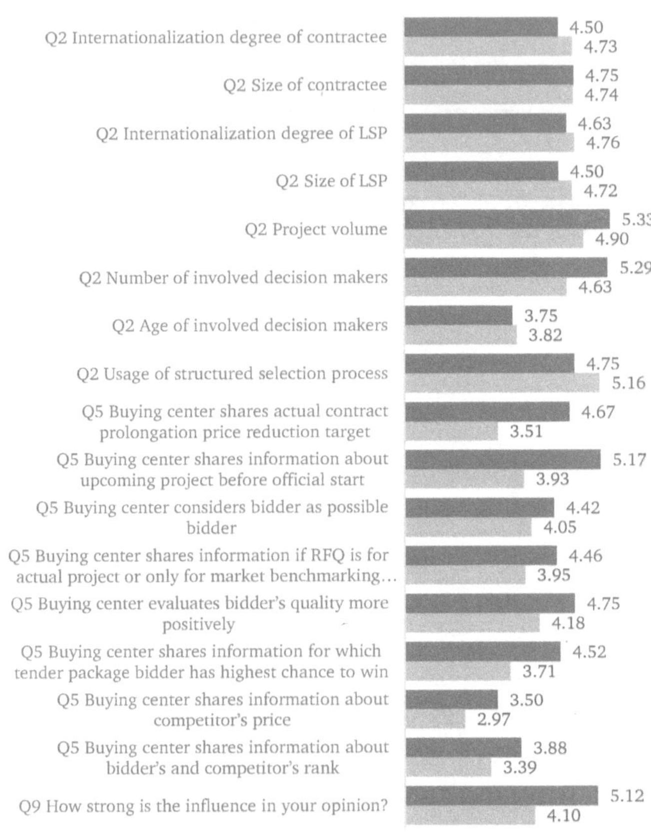

Figure 23: Means of personal relationship influence questions (Q2, Q5, Q9) for LSPs compared to Buying companies
Source: Own figure

Females compared to Males

Responses from female and male participants show no considerable discrepancy with a mean value difference of just 0.05 across all questions Q2, Q5, and Q9 (females: 4.20, males: 4.25). In Q2, females' mean value is 4.70, while for males, it is 4.65, showing a slightly higher value of 0.05 for females. In contrast, for Q5, the outcome is reversed with a mean value for males with 3.86 being 0.16 higher than for females with 3.7. The results are shown in Table 11.

Table 11: Differences of personal relationship influence by gender

Group	Q2	Q5	Q9	Total	
Females	4.70	3.70	4.20	4.20	
Males	4.65	3.85	4.24	4.25	Δ 0.05
Total	4.66	3.77	4.23	4.22	

This high similarity is also observable when comparing means for the factors of the related questions. For eight factors, the difference is below 0.10, such as for *Number of involved decision makers*, where it is zero. Males rate *Buying center considers bidder as possible bidder* and *Buying center shares information for which tender package bidder has highest chance to win* slightly higher with a difference of 0.29 and 0.31, respectively. On the other hand, females rate *Usage of structured selection process* higher than males with a value of 5.22 compared to 4.98. The results are shown in Figure 24.

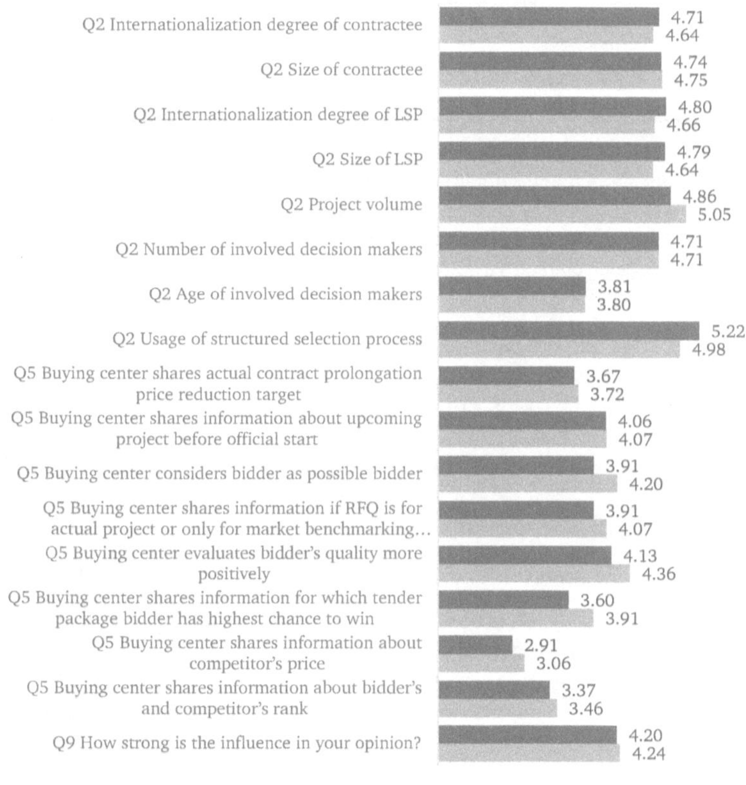

Figure 24: Means of personal relationship influence questions (Q2, Q5, Q9) for Females compared to Males
Source: Own figure

Little compared to Medium compared to Much work experience

Comparing the three different clusters of years of work experience, a certain, albeit small, trend can be observed: Less work experience leads to higher mean values. While for the group of little work experience, the mean value across all questions is 4.33, it is 4.20 for the group with medium work experience and 4.11 for the group with much work experience. This can be seen in Table 12.

Table 12: Differences of personal relationship influence by work experience

Group	Q2	Q5	Q9	Total	
Little work experience	4.81	3.83	4.48	4.33	Δ 0.13
Medium work experience	4.68	3.73	4.20	4.20	Δ 0.09
Much work experience	4.51	3.73	3.96	4.11	
Total	4.66	3.77	4.23	4.22	

In 12 out of 17 factors, participants with little work experience reached the highest mean value. The most considerable differences between each cluster occur at the factors *Age of involved decision makers* (Δ 0.30; Δ 0.39) and *How strong is the influence of personal relationships on the LSP selection process in your opinion* (Δ 0.28; Δ 0.24). Figure 25 presents the breakdown of the mean values for all three clusters.

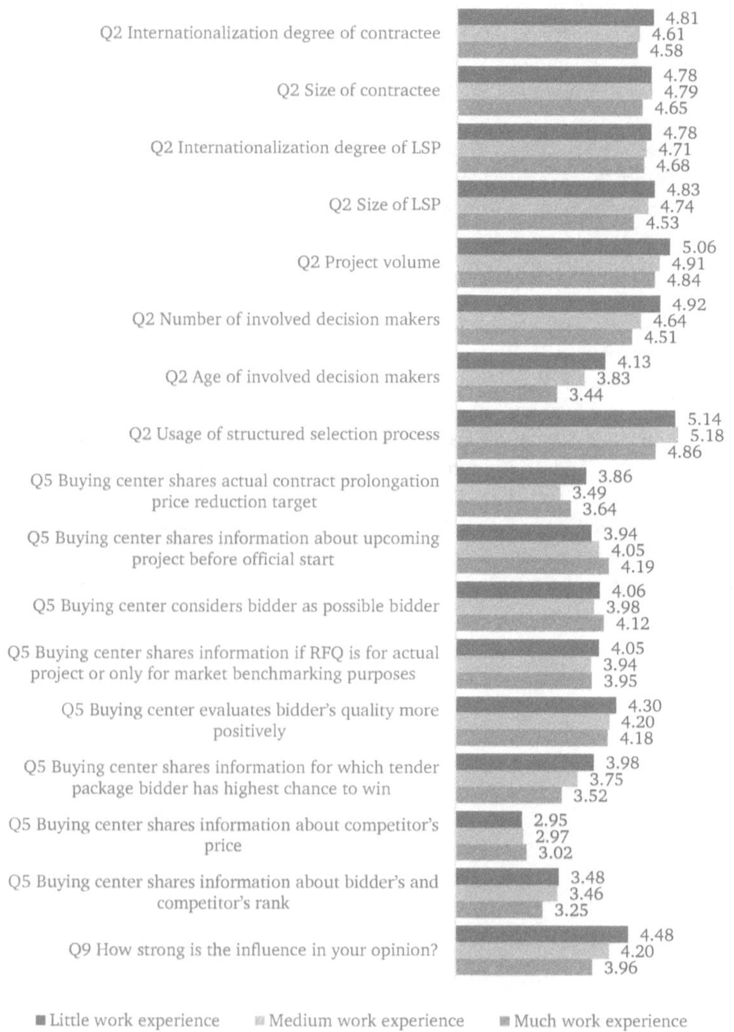

Figure 25: Means of personal relationship influence questions (Q2, Q5, Q9) based on work experience
Source: Own figure

6.3.3. Mann-Whitney-U-Test

The MWU-Test is used to confirm the research propositions by considering frequencies. The demographic questions from Q10, Q11, Q12, Q13, Q14, and Q16 act as independent variables; the answers to Q2, Q5, and Q9 form the dependent ones.

Table 13 shows the result of the MWU-Test for the opposing pairs *China – Western Society, Small and medium sized companies – Large enterprises, Multinational enterprises – Domestic companies* and *Logistics Service Providers – Buying companies*. The pairs *Females – Males* and *Little work experience – Much work experience* have also been tested, but as no significant difference between the groups could be observed, the results are not presented. Comparing *SMEs* and *Large enterprises* reveal significant results in different directions. SMEs reach a lower mean rank in one case and a higher one in the other. For five questions, the mean ranks of the answers from *LSPs* are significantly higher than those of *Buying companies*. For 10, respectively 11 questions, the answers are consistently significantly higher for *China* versus *Western Society* and *Domestic company* versus *Multinational enterprise*. According to the previous section's findings, the frequency distribution and the mean values indicate that both the country in which the participants are employed and whether working in multinational or domestic companies have a significant influence. Responses are consistently higher for *China* and *Domestic companies* than for *Western Societies* and *Multinational enterprises*. This observation is also confirmed by the MWU-Test, where a significant difference between *China* and *Western Society* as well as *Multinational enterprises* and *Domestic companies* can be demonstrated. Similarly, even though in limited form, this is true for *LSPs* and *Buying companies*, which might be caused by the uneven sample distribution with 24 LSPs and 154 Buying companies. Table 13 shows the results of the MWU-Test. The results, including U-values and Z-values, are provided in Appendix s2.

Table 13: Survey MWU-Test results

	Mean rank		Sig.	Mean rank		Sig.	Mean rank		Sig.	Mean rank		Sig.
	China	Western Societies	r	SMEs	Large enterprises	r	Multi-national companies	Domestic companies	r	LSPs	Buying companies	r
Q2 Internationalization degree of contractee	84.20	55.91	0.000	87.07	96.22	0.528	82.38	114.75	0.000	84.58	90.27	0.608
Q2 Size of contractee	83.64	56.91	0.000	94.57	95.58	0.944	82.94	113.93	0.000	88.38	89.68	0.907
Q2 Internationalization degree of LSP	82.84	58.33	0.001	100.97	94.49	0.654	80.68	115.83	0.000	85.44	89.56	0.709
Q2 Size of LSP	82.82	58.35	0.001	79.07	96.91	0.218	83.78	112.69	0.000	80.94	90.83	0.371
Q2 Project volume	77.15	68.41	0.220	104.73	94.71	0.489	86.35	108.92	0.005	97.35	88.28	0.412
Q2 Number of involved decision makers	72.27	72.93	0.926	65.50	95.78	0.039	92.79	94.53	0.824	106.52	84.46	0.042
Q2 Age of involved decision makers	81.01	59.93	0.003	86.07	95.77	0.502	90.36	101.75	0.153	85.25	89.59	0.695
Q2 Usage of structured selection process	82.47	55.55	0.000	80.53	95.71	0.290	82.96	111.13	0.000	79.48	89.92	0.340
Q5 Buying center shares actual contract prolongation price reduction target	75.01	67.69	0.315	106.82	92.96	0.350	89.41	100.70	0.155	119.92	83.54	0.001
Q5 Buying center shares information about upcoming project before official start	74.75	69.77	0.487	109.18	93.32	0.286	89.18	102.34	0.098	122.04	83.82	0.001
Q5 Buying center considers bidder as possible bidder	76.23	65.48	0.133	76.46	95.42	0.199	89.26	101.07	0.136	98.92	87.44	0.299
Q5 Buying center shares information if RFQ is for actual project or only for market benchmarking purposes	76.39	65.19	0.117	96.39	93.81	0.861	85.91	105.81	0.012	103.94	86.06	0.104
Q5 Buying center evaluates bidder' s quality more positively	78.87	62.18	0.020	90.32	94.30	0.787	91.73	97.32	0.479	105.71	85.78	0.069
Q5 Buying center shares information for which tender package bidder has highest chance to win	79.51	59.32	0.005	126.08	91.25	0.028	84.17	107.31	0.004	109.09	85.41	0.035
Q5 Buying center shares information about competitor' s price	82.84	54.86	0.000	104.32	93.71	0.472	83.81	110.25	0.001	101.06	87.11	0.204
Q5 Buying center shares information about bidder' s and competitor' s rank	82.38	55.72	0.000	101.07	93.97	0.634	86.08	106.91	0.009	102.33	89.91	0.165
Q9 How strong is the influence in your opinion?	78.28	64.86	0.060	108.93	93.80	0.294	84.43	110.38	0.001	116.10	84.75	0.004

6.3.4. Explorative Factor Analysis

The EFA is used to illuminate to what extent the response structure of individual variables is similar. Such variables can then be combined, forming a common factor. Answers to the questions Q2, Q5, and Q9 are selected as input for the EFA, as they aim to quantify the influence of personal relationships on the selection process. The correlation and significance matrices are shown in Table 14. Significant values in the second part of the table confirm the findings at a significance level $p < 0.05$.

Table 14: Correlation and significance matrix for EFA

		Q2 1	Q2 2	Q2 3	Q2 4	Q2 5	Q2 6	Q2 7	Q2 8	Q5 1	Q5 2	Q5 3	Q5 4	Q5 5	Q5 6	Q5 7	Q5 8	Q9
Correlation	Q2 1	1.000																
	Q2 2	0.678	1.000															
	Q2 3	0.735	0.636	1.000														
	Q2 4	0.573	0.604	0.704	1.000													
	Q2 5	0.502	0.538	0.469	0.526	1.000												
	Q2 6	0.314	0.337	0.285	0.316	0.542	1.000											
	Q2 7	0.407	0.431	0.366	0.431	0.421	0.464	1.000										
	Q2 8	0.538	0.513	0.498	0.532	0.49	0.436	0.473	1.000									
	Q5 1	0.081	0.16	0.107	0.113	0.192	0.305	0.115	0.007	1.000								
	Q5 2	0.124	0.058	0.126	0.059	0.271	0.264	0.060	0.056	0.628	1.000							
	Q5 3	0.132	0.106	0.164	0.103	0.184	0.235	0.093	0.223	0.504	0.56	1.000						
	Q5 4	0.188	0.189	0.201	0.144	0.33	0.219	0.116	0.205	0.548	0.565	0.591	1.000					
	Q5 5	0.232	0.212	0.282	0.142	0.265	0.254	0.134	0.166	0.462	0.485	0.544	0.554	1.000				
	Q5 6	0.21	0.235	0.231	0.133	0.386	0.224	0.209	0.114	0.503	0.531	0.481	0.556	0.657	1.000			
	Q5 7	0.201	0.233	0.268	0.14	0.215	0.231	0.271	0.178	0.451	0.345	0.392	0.419	0.476	0.650	1.000		
	Q5 8	0.142	0.269	0.242	0.175	0.292	0.343	0.186	0.225	0.500	0.362	0.399	0.519	0.456	0.616	0.754	1.000	
	Q9	0.284	0.239	0.272	0.18	0.373	0.331	0.207	0.144	0.445	0.333	0.336	0.351	0.254	0.401	0.310	0.380	1.000
Significance (1-tailed)	Q2 1																	
	Q2 2	0.000																
	Q2 3	0.000	0.000															
	Q2 4	0.000	0.000	0.000														
	Q2 5	0.000	0.000	0.000	0.000													
	Q2 6	0.000	0.000	0.000	0.000	0.000												
	Q2 7	0.000	0.000	0.000	0.000	0.000	0.000											
	Q2 8	0.000	0.000	0.000	0.000	0.000	0.000	0.000										
	Q5 1	0.141	0.017	0.077	0.066	0.005	0.000	0.064	0.465									
	Q5 2	0.051	0.221	0.047	0.217	0.000	0.000	0.215	0.231	0.000								
	Q5 3	0.040	0.081	0.015	0.087	0.007	0.001	0.108	0.001	0.000	0.000							
	Q5 4	0.006	0.006	0.004	0.028	0.000	0.002	0.062	0.003	0.000	0.000	0.000						
	Q5 5	0.001	0.002	0.000	0.03	0.000	0.000	0.038	0.014	0.000	0.000	0.000	0.000					
	Q5 6	0.003	0.001	0.001	0.039	0.000	0.001	0.003	0.065	0.000	0.000	0.000	0.000	0.000				
	Q5 7	0.004	0.001	0.000	0.032	0.002	0.001	0.000	0.009	0.000	0.000	0.000	0.000	0.000	0.000			
	Q5 8	0.030	0.000	0.001	0.010	0.000	0.000	0.007	0.001	0.000	0.000	0.000	0.000	0.000	0.000	0.000		
	Q9	0.000	0.001	0.000	0.008	0.000	0.000	0.003	0.028	0.000	0.000	0.000	0.000	0.000	0.000	0.000	0.000	

The significant matrix is symmetric and thus confirms the suitability of the data for an EFA. This is also valid for Bartlett's Test of Sphericity at a significance level of 0.000 (Approx. Chi-Square: 1654.797, df: 136). According to the KMO MSA, the suitability of the data for an EFA can be considered *meritorious* (KMO 0.854)[881]. Hence, the EFA could be conducted. The results are shown in Table 15. Four factors have eigenvalues greater than 1 so that according to the Kaiser-criterion, a total of four factors should be determined. Factor 1 explains 37.3% of the total variance, factor 2 18.6%, and factors 3 and 4, each slightly more than 6%. Cumulatively, these four factors explain 68.1% of the variance.

Table 15: Explained variance of EFA

Factor	Initial Eigenvalues	Explained variance	Cumulative explained variance
1	6.344	37.317	37.317
2	3.158	18.577	55.894
3	1.046	6.153	62.048
4	1.022	6.011	68.059
...			
17	0.171	1.008	100.000

The scree plot is shown in Figure 26. Its elbow, the bend where the difference of the eigenvalues between two factors is greatest, is visible at factor 3. Following the elbow criterion, the number of factors extracted should, therefore, be two.

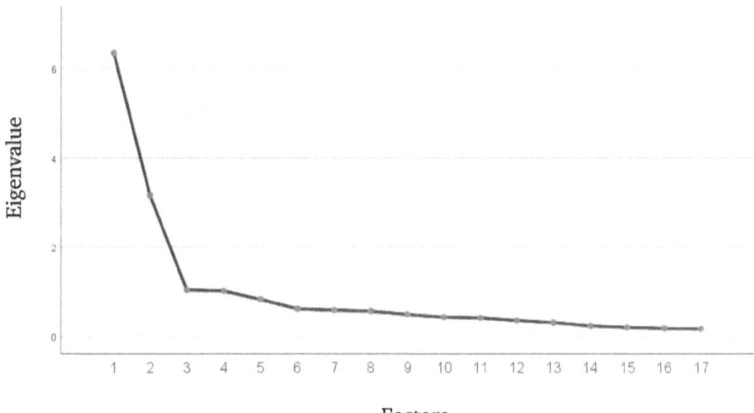

Figure 26: Scree plot of EFA
Source: Own figure

As the number of factors that should be determined differs based on the results of the Kaiser criterion and the scree plot, a decision is to be made according to content interpretation. This can be done based on the rotated component matrix, shown in Table 16. Initially, four factors are considered. The respective highest factor loading of each variable is colored in dark grey. If a variable cannot be clearly assigned to one factor, as the values are either too low or similarly high on two factors, they are colored in light grey. A total of five variables can be uniquely assigned to factor 1 and five variables uniquely assigned to factor 2. Two further variables each have a value greater than 0.5 for factors 1 and 2, but also a somewhat high value for the alternative factors 3 and 4.

Table 16: Rotated component matrix for EFA

	Component			
	1	2	3	4
Q2 Internationalization degree of contractee	0.851	0.114	0.031	0.106
Q2 Size of contractee	0.791	0.043	0.168	0.187
Q2 Internationalization degree of LSP	0.872	0.122	0.131	0.016
Q2 Size of LSP	0.806	0.04	0.013	0.184
Q2 Project volume	0.541	0.255	0.043	0.529
Q2 Number of involved decision makers	0.212	0.212	0.070	0.819
Q2 Age of involved decision makers	0.404	-0.090	0.230	0.620
Q2 Usage of structured selection process	0.647	0.026	0.050	0.382
Q5 Buying center shares actual contract prolongation price reduction target	-0.051	0.725	0.256	0.218
Q5 Buying center shares information about upcoming project before official start	-0.008	0.839	0.049	0.126
Q5 Buying center considers bidder as possible bidder	0.086	0.773	0.140	0.039
Q5 Buying center shares information if RFQ is for actual project or only for market benchmarking purposes	0.143	0.758	0.245	0.045
Q5 Buying center evaluates bidder's quality more positively	0.212	0.647	0.389	-0.073
Q5 Buying center shares information for which tender package bidder has highest chance to win	0.133	0.574	0.614	0.068
Q5 Buying center shares information about competitor's price	0.117	0.281	0.872	0.101
Q5 Buying center shares information about bidder's and competitor's rank	0.089	0.352	0.788	0.215
Q9 How strong is the influence of personal relationships on the selection of LSPs in your opinion?	0.134	0.458	0.141	0.408

Based on the rotated factor matrix, the factor allocation shown in Table 17 can be derived. The order is according to the size of the loading value, and variables that do not clearly load on one factor are listed in brackets. Q9 is not further considered, as the factor loadings of Q9 do not reach the predefined minimum of 0.50.

Table 17: Factor classification

Factor	Variables
1	Q2 Internationalization degree of LSP
	Q2 Internationalization degree of contractee
	Q2 Size of LSP
	Q2 Size of contractee
	Q2 Usage of structured selection process
	(Q2 Project Volume)
2	Q5 Buying center shares information about upcoming project before official start
	Q5 Buying center considers bidder as possible bidder
	Q5 Buying center shares information if RFQ is for actual project or only for market benchmarking purposes

		Q5 Buying center shares actual contract prolongation price reduction target
		Q5 Buying center evaluates bidder's quality more positively
	3	Q5 Buying center shares information about competitor's price
		Q5 Buying center shares information about bidder's and competitor's rank
		(Q5 Buying center shares information for which tender package bidder has highest chance to win)
	4	Q2 Number of involved decision makers
		Q2 Age of involved decision makers

The factor allocation mostly follows the original questions, so that variables from Q2 (factors 1 and 4) and variables from Q5 (factors 2 and 3) can form independent factors.

Factor 1 mainly includes company-related variables, describing both the contractor and contractee. Therefore, Factor 1 is named *"Company characteristics that influence the impact of personal relationships on the LSP selection."* Factor 2 comprises five variables concerning various support possibilities that the buying center can offer to help an LSP during the selection process. Therefore, Factor 2 is named *"Support possibilities of a Buying center representative to help a certain LSP during the selection process."* Factor 3 is formed of the other two (respectively three) variables, queried in Q5. They are variables that relate to particularly commercially sensitive information about the LSP's competitors and its position related to them. Factor 3 is therefore named *"Commercially sensitive information a Buying center representative can share to help a certain LSP during the selection process."* The last factor comprises two variables with attributes related to the buying center in terms of personal characteristics and setup. Following the wording of Factor 1, Factor 4 is named *"Buying center characteristics that influence the impact of personal relationships on the LSP selection."* Factor values can then be calculated for all identified factors and used as the basis for regression analysis.

6.3.5. Regression Analysis

Twelve independent variables were selected for the RA. Based on the elbow criterion, factor values of the Factors 1 and 2 that were determined in the EFA are considered in the regression analysis as dependent variables. Besides, Factor 3 is considered due to the strong result of the frequency distribution with a significantly lower mean value than the other possible answers in Q5. Therefore, a total of 36 simple regression analyses were carried out. The results are shown in Table 18. Out of these 36 RAs, 11 have a significance level of $p < 0.05$, highlighted in grey.

Table 18: Regression Analysis results

Independent Variables	Factor 1			Factor 2			Factor 3		
	Coeff.	Sig.	R^2	Coeff.	Sig.	R^2	Coeff.	Sig.	R^2
Q8 Structured Selection Process	.270	.611	.010	.125	.016	.033	.010	.845	.000
Q10 Firm Size small	-.127	.684	.010	.072	.818	.000	.479	.125	.013
Q10 Firm Size large	.127	.684	.010	-.072	.818	.000	-.479	.125	.013
Q11 Multinational Enterprise	-.670	.000	.110	-.114	.457	.003	-.346	.023	.029
Q11 Domestic Company	.670	.000	.110	.114	.457	.003	.346	.023	.029
Q12 LSP	-.335	.137	.130	.728	.001	.061	.087	.700	.010

Q12 Automotive Industry	.347	.081	.170	-.477	.016	.033	.048	.809	.000
Q13 Female	.860	.574	.002	-.016	.917	.000	-.110	.478	.003
Q13 Male	-.860	.574	.002	.016	.917	.000	.110	.478	.003
Q14 China	.581	.000	.084	.092	.542	.002	.536	.000	.072
Q14 Western Society	-.558	.001	.058	.019	.913	.000	-.625	.000	.072
Q16 Years of work experience	.003	.799	.000	-.003	.794	.000	-.007	.499	.030

Q8 – Structured Selection Process. The usage of a structured selection process is positively correlated with all three factors. The influence of personal relationships on the support possibilities will increase if the selection process is more structured, which contradicts the original research proposition. However, this positive correlation is only significant for Factor 2.

Q10 – Firm Size. No clear direction of a firm size effect is observable, and the analysis results are not significant. Noteworthy are the results at Factor 3, where a comparatively high regression coefficient of 0.479 and a relatively low level of significance of 0.125 suggest that the willingness to pass on sensitive information is greater in smaller companies than in large companies. The low significance level is likely caused by the small sample size of 15 respondents from SMEs.

Q11 – Internationalization Degree. Multinational enterprises have negative regression coefficients for all three factors, while domestic companies have positive ones. The results are significant for both Factor 1 and Factor 3. Hence, employees from domestic companies rate the influence of personal relationships higher and are more willing to provide information concerning an LSP's competitor than their counterparts from multinational enterprises.

Q12 – Firm Type. The firm type LSP is significantly positively correlated with Factor 2. The regression coefficient of 0.728 is the highest significant value among all conducted regression analyses. Factors 1 and 3 are neither significant, nor do they have the same effect direction.

Q13 – Sex. Neither a uniform direction of effect nor a significant result can be observed.

Q14 – Country of Work. Comparing China and Western Societies, the China cluster consistently correlates positively, and the Western Societies cluster consistently negatively with all factors. Like "internationalization degree," the regression coefficients of Factor 1 and 3 are significant and relatively high. Employees working in China rate the influence of personal relationships higher than their counterparts in Western Societies, and are also more willing to provide sensitive information.

Q16 – Years of Work Experience. Neither a uniform direction of effect nor significant results are observable.

6.4. Discussion

In the following section, the multilayered results of the study are discussed. First, the research outcomes are summarized and evaluated. Then limitations and implications are pointed out, and recommendations for future research are given.

Table 19: Result RQs1

6.4.2. Evaluation

The findings of this study shed light on the support possibilities during the selection process for LSPs, as well as contingency factors that may alter the support possibilities. They will be related to current research and the outcomes from the explorative study described in chapter 5.

In *RQs1*, it was asked which contingency factors influence the impact of personal relationships on the LSP selection. Eight research propositions were developed that could be partially confirmed. In *RPs1*, it was proposed that if a decision maker is younger, the influence of personal relationships on the LSP selection process is lower than if the decision maker is older. However, this research proposition could not be confirmed. Although most respondents assumed that this phenomenon is indeed correct, younger survey participants actually selected higher values than older ones. It might be explained by the assimilation of the organizational culture of employees who are working longer in a company environment, so that they are more aware of possible compliance-related impacts of their actions. Values and related behavior not only diverge throughout different generations but also across different organizations[882]. In this business organizational context, values are "evaluative standards relating to work or the work environment by which individuals discern what is 'right' or assess the importance of preferences[883]." They are learned through direct experience or influence processes, and their influence is increasing with more knowledge about the object of the value[884]. Under the assumption that companies are moving towards compliant business practices and are conducting corresponding training, older employees have already been stronger impacted by such organizational values and behave accordingly[885]. *RPs2* states that the sex of the boundary spanners has no influence on the impact of personal relationships on the LSP selection process. This research proposition could be confirmed throughout the study. It aligns with viewpoints according to which men and women's behavior at work is more similar than different[886]. In *RPs3*, it was proposed that the influence of personal relationships on the LSP selection process is lower in Western cultures compared to China. This research proposition could also be confirmed. Apparently, bonds among individuals in China transcend organizational boundaries differently than in the West[887]. *RPs4* brought the notion forth that firm size impacts the influence of personal relationships on the LSP selection. It was presumed that at large companies, the influence is lower than in SMEs. The study results tend to confirm this proposition. Employees at SMEs seem to employ personal relationships more extensively than their counterparts at large companies. *RPs5* addresses a company's internationalization degree, stating that at multinational companies, the influence of personal relationships on the LSP selection process is lower than in domestic companies. This research proposition could also be confirmed. Companies engaged in international operations accommodate their behavior towards international business conduct that seeks rather to limit the impact of personal relationships in tender processes[888]. According to the equally confirmed *RPs6*, LSPs perceive the influence of personal relationships on the LSP selection process higher than buying firms. While this seems somewhat surprising, considering that both LSPs and Buying firms are in a dyadic relationship during a tender, various explanations are possible. Associates working at LSPs might have a

---

[882] Inglehart (2008), p. 130; Loughlin/Barling (2001), p. 543; Finegan (2000), pp. 149-155.

[883] Dose (1997), pp. 227-228.

[884] Fazio/Zanna (1981), pp. 161-198.

[885] Josephson (2014), pp. 13-15; Weber/Fortun (2005), p. 97; Thomson Reuters (2016).

[886] Eagly/Johannesen-Schmidt (2001), pp. 781-782; Heilman (2012), pp. 123-131; Hopkins et al. (2008), pp. 348-349.

[887] Jia et al. (2016), pp. 1249-1250; Wang (2007), pp. 81-85; Davies et al. (1995), pp. 208-213.

[888] Keefer/Knack (1997), pp. 591-594; Asgary/Mitschow (2002), pp. 240-245; Schwartz (2005), pp. 36-39; He/Cui (2012), pp. 352-355.

larger fundus of experience with various clients or other associates from a buying company so that they have more profound insights into the actual impact of personal relationships. In contrast, associates at a buying firm comparatively underestimate it. An indication for that could be that they made, on average, 5.6 statements per person, while Buying firms' associates made only 3.6 during the explorative expert interviews. Alternatively, LSP employees might perceive the impact of personal relationships higher than it actually is. Members of a selling center are incentivized to set up close personal relationships as it can have certain sales benefits so that they likely do so[889]. Following self-perception theory, people develop their attitudes under certain circumstances by observing their own behavior and concluding that certain attitudes must have caused it[890]. Likewise, cognitive dissonance theory provides an explanation. When a person participates in an action that goes against a held belief or idea, uncomfortable tension occurs. Therefore, a person will work on reducing this tension by, for instance, changing the belief until it is consistent with the action[891]. Either theory explains why key account managers would believe that personal behavior strongly impacts the selection process. An indication for this could be the statement from the Key account manager Expert B in the qualitative study who explains that "we must know the people at the customers (...) that's why companies should have a sales. The salespeople should open the door to this company at first. The policy or KPI is the second part, but the sales needs to open the door (...) Guanxi is only for ice broke because I need to know the people." *RPs7*, which states that in case of a large project volume, the influence of personal relationships on the LSP selection process is lower than in case of a small project volume, could not be verified due to a lack of data. Also, respondents contradicted the proposition by assuming that in the case of large project volumes, the influence of personal relationships on the LSP selection process is higher than in the case of small project volumes. A reason for this could be that boundary spanners try to use their personal relationships only for projects that have a high reward potential. For instance, as LSPs' key account managers only have limited available resources, they must concentrate their efforts - and likely do so - on big volume projects[892]. According to *RPs8*, if a large number of decision makers are involved, the influence of personal relationships on the LSP selection process is lower than if only a few decision makers are involved. This proposition could also not be verified due to insufficient data. A reason for a less clear outcome could be that with an increased number of decision makers, also the numbers of people that might be influenced due to personal relationships increases. Likewise, *RPs9* could not be confirmed. It suggests that if a highly structured LSP selection process is used, the influence of personal relationships on the LSP selection process is lower than if an unstructured selection process is used. However, results from Q3 and the EFA/RA show that the influence of personal relationships is greater if the selection process is more structured. While this seems counterintuitive, it might be explicable due to the nature of structured processes. They are formally defined, standardized, and can be identically repeated[893]. This might increase the possibilities for the influence of personal relationships as it is more obvious which process steps will happen, and it could be more transparent how and when personal relationships might positively impact the selection.

RQs2 addressed how members of a buying center support a certain LSP during the selection process. Eight Support Possibilities identified in the qualitative study were tested, and several exciting findings were generated. Two of the top three most frequently occurring support

[889] Millman/Wilson (1995), p. 13.
[890] Bem (1972), pp. 1-62.
[891] Cooper (2007), pp. 1-180; Festinger (2009), pp. 1-176.
[892] Campbell/Cunningham (1983), pp. 369-380.
[893] Kroenke/Boyle (2017), p. 286; Schäfermeyer et al. (2010), p. 2.

possibilities are SP2, *Buying center shares information about upcoming project before official start* and SP3, *Buying center considers bidder as possible bidder*, which are both relevant before a tender actually starts. The least frequent support possibilities, such as SP7, *Buying center shares information about competitor's price* and SP8, *Buying center shares information about bidder's and competitor's rank*, occur at later stages in the LSP selection process. This might be related to the difficulty of detecting manipulation or such leaks beforehand[894]. An alternative explanation could be that associates are not aware of a possibly harmful impact on the selection process through, e.g., sharing information about upcoming projects before the official start. Regarding the support possibilities related to the sharing of information, SP1, SP2, SP4, SP6, SP7, and SP8, it can be assumed that sharing is reduced with increasing commercial sensitivity[895]. For instance, highly sensitive information, such as competitors' prices (SP7) or bidder's and competitor's ranks (SP8), is shared least frequently. Differences in the sharing of sensitive information are also visible comparing Western societies and China. In Western societies, highly sensitive information such as competitors' prices (SP7) is almost never shared (value 2.27, the lowest value among all support possibilities). In China, this information is shared at an almost as high rate (3.47) as information in Western societies about the actual contract prolongation price reduction target (3.56).

6.4.3. Limitations, implications and future research

In this section, the limitations and implications of this study are stated, as well as suggestions for future research suggested.

Limitations

In this study, a global survey was conducted among supply chain experts to address the gap of the qualitative study. Naturally, approaches of this kind carry with them limitations.
Some propositions could not be verified due to missing data. This is caused by the necessary tradeoffs when designing survey instruments. Research objects should be specified and delimited very precisely, but questionnaires can practically only be of a certain length to avoid discouraging survey respondents by an overwhelming set of questions. This study's target was to generate holistic insights, inquired through an extensive number of contingency factors and support possibilities, which resulted in substantial question complexity. Therefore, it was decided, following common recommendations about questionnaire design, to take the risk of not being able to answer certain parts of the research question in favor of respecting an upper limit for the questionnaire length[896]. Hence, also contingency factors were only generally confirmed instead of strengths of the impact of various contingency factors on specific support possibilities measured. Also, sampling was not done as probability sampling but instead as nonprobability sampling, where samples from a larger population were drawn subjectively instead of randomly selected[897]. While this is a commonly used method in social science research and was chosen deliberately given the total population characteristics, the study results cannot be unrestrictedly projected to the whole population[898]. Moreover, the sample size was partly imbalanced, with, for example, 15 participants working for SMEs, while 175 work for large enterprises. As a result, some outcomes of the study have to be cautiously interpreted.

---

[894] Porter/Zona (1993), pp. 518-538.
[895] Rosenblum/Maples (2009), pp. 33-34.
[896] Döring et al. (2015), p. 411.
[897] Tansey (2007), pp. 768-770.
[898] Feild et al. (2006), pp. 566-568; Tansey (2007), pp. 768-770.

Furthermore, the study was conducted mostly in the automotive industry, so that, despite this sector's comprehensive scope and impact, the results may not be generalized to all industries.

Implications

In spite of these limitations, the findings have important implications. From a *theory* view, first, it extends research about contingency factors that mitigate the Guanxi impact. Luo et al. identified ownership structure, state-owned vs. non-state-owned organizations, location, mainland vs. overseas China, and time[899]. Based on the qualitative study (chapter 5), it can be contended that support possibilities are altered based on certain contractor/contractee characteristics (internationalization degree, size), project characteristics (project volume, number of involved decision makers, usage of structured selection process) and personal characteristics (age, sex, home country). In this subsequent study, these contingency factors were reassessed quantitatively. Internationalization degree of the contractor/contractee could be confirmed. This supports theses by researchers that assume that businesses' internationalization of will lead to a convergence of business practices[900]. As some of the practices discussed in the study during tender processes might be perceived as violating international laws and codes of conduct, companies that are following international business procedures refrain from using it[901]. Likewise, the size of the contractor/contractee was verified. It endorses the notion that smaller companies are more likely to violate anti-corruption regulations than larger companies, considering that its employees enjoy a high degree of autonomy and are seldom subject to checks by their colleagues[902]. Ample of evidence also indicated differences between decision makers' home countries. Despite the internationalization of business practices, cultural differences in the workplace seem to be still valid. In rather collectivist countries, relationships have more the character of a family relationship compared to the business transaction style of rather individualist countries[903]. Regarding age, no significant difference could be found between older and younger decision makers. Interestingly, this contradicts assumptions that younger generations consider relationship-based work practices as outdated and thus behave differently from their older peers[904]. Also, no difference was found for the attribute "sex," which is not in line with studies that show gender differences about social capital and communication in the workplace due to, e.g., gender stereotypes, conferred legitimacy and status[905]. Second, the study contributes to the understanding of Guanxi practices on a micro level within supply chain management. It answers the call of, e.g., Zhang and Zhang, to extend the research scope to not only incorporate social ties of top executives but also among employees at "lower levels[906]." It was revealed that employees of a buying center especially share information about upcoming projects before the official start, evaluate a bidder's quality more positively, or consider bidders as possible bidder. This also broadens the understanding of the Guanxi impact on supply chain management. While a link to risk management, strategic purchasing, or green supply chain management could be demonstrated, its impact on the selection of LSPs, an increasingly relevant topic in China, was

---

[899] Luo et al. (2012), p. 148.
[900] Asgary/Walle (2002), pp. 69-70.
[901] Asgary/Mitschow (2002), p. 240; Lindgreen (2004), pp. 35-37.
[902] Wells (2014), pp. 41-42; Linnenluecke/Griffiths (2010), pp. 358-360.
[903] Hofstede (2001), 119-123.
[904] Chen et al. (2013), pp. 193-194; Wilson/Brennan (2010), pp. 662-663; Faure/Fang (2008), p. 197; Hanser (2002), pp. 160-161.
[905] Timberlake Sharon (2005), pp. 41-43.
[906] Zhang/Zhang (2006), pp. 389-390.

so far not addressed[907]. The study further adds to the literature about BSRs during the procurement of services which are defined by intangibility and heterogeneity[908]. Guanxi can positively impact the evaluation of an LSP's quality during the critical evaluation stage. It also potentially leads to an information advantage that might help potential contractors to improve their offering irregularly. This research also shows the limitations of personal relationships. Seldom, buying center employees share information of high commercial sensitivity such as LSP competitors' prices or bidders' rank. Also, from a *practical* point of view, this study has profound implications for contractees of logistics services as well as for LSPs. While differences across regions can be observed, overall, there is consensus that personal relationships profoundly impact the selection of Logistics Service Providers. Acknowledging that this effect exists and that this might have a certain influence is the first step to making a deliberate decision to either accept or develop measures to avoid, control or transfer it[909]. Ignoring such effects poses the risk of various negative consequences. Commercially and operationally, buying companies could suffer from selecting an LSP with inflated prices or sub-par day-to-day operations. Besides, there is reputational risk in the industry. If a contractee is known for its strong personal relationship influenced business practices, capable LSPs might refrain from making proper tender offers as their costs of doing so would outweigh the benefits in the form of a realistic chance to obtain the contract[910]. Especially companies operating in various regions should be aware of local geographical particularities with varying relationship patterns and different relationship strengths[911]. Companies might consider setting up procedures that minimize the chance for any impact of personal relationships. This could include a mandatory change of personnel within a certain time or the use of automated quality evaluations. That sharing of highly commercially sensitive information is a rather seldom occurrence can indicate that buying center employees are alert about the relevance and possible implications of such behavior and that relevant compliance measures are effective. On the other hand, it seems that this awareness is not extended yet towards behavior that occurs before a tender start. Improving the alertness at what stages the risk of bid-rigging already begins, respectively, implementing relevant mitigation measures could lead to fairer tender processes. This study also indicates that personal relationships have a higher impact on certain types of firms. Companies that are small and medium-sized, of Chinese origin and mainly operating domestically, are exposed to a higher risk of personal relationships than their opposing counterparts. If a company falls into one of these categories, it must be particularly watchful about possible negative impacts through personal relationships and react accordingly. For instance, small and medium-sized companies with traditionally lower budgets than large enterprises, should invest in measures that ensure regulatory compliance, such as training and oversight. Previous research has established that diversity in organizations can have various positive effects, such as improved decision making with increased creativity and innovation[912]. However, in this study, no difference in decision maker characteristics could be reported. Younger decision makers compared to older decision makers and female compared to male decision makers seem to be equally susceptible to the impact of personal relationships in tender processes. Setting up a diverse buying center might therefore not influence compliance adherence. Consequently, this study has also interesting

---

[907] Abramson/Ai (1999), pp. 21-31; Cheng et al. (2012), p. 11; Lee/Humphreys (2007), pp. 462-463; Lee/Humphreys (2007), pp. 454-462; Zhu/Sarkis (2004), pp. 282-285; Simpson et al. (2007), pp. 42-44; Luo et al. (2014), pp. 105-107; Geng et al. (2017), p. 3.

[908] Zeithaml et al. (2018), 6-7; Jackson et al. (1995), pp. 100-106.

[909] Garvey (2009), pp. 2-10.

[910] United Nations Office on Drugs and Crime (2013), pp. 10-11.

[911] Clark (1990), pp. 66-79.

[912] Roberge/van Dick (2010), pp. 297-298.

implications for Logistics Service Providers. Personal relationships are an important element when contractees decide for an LSP. Therefore, key account managers could enhance it by using Guanxi practices, such as fostering a sense of friendship with their clients in the form of gift-giving or client gatherings. LSPs should provide training and support to their salespeople so that they can enhance their skills in developing personal relationships, which are connected to increased business success[913]. LSPs could also identify current and future members of buying centers and investigate if somebody in their own organization possesses related Guanxi with such members, respectively, can build it promptly. Alternatively, LSPs could hire certain employees that can contribute the required resources. As Guanxi relationships may become stale and need rejuvenation, such internal "Guanxi audits" should be repeated regularly, for instance, matching the contract renewal cycles of contractees[914]. Considering that buying centers most likely share information in the early phases of a tender or even before the official start of it, LSPs should focus their sales efforts early on. For example, they could obtain important information such as when a project will begin, granting them crucial advantages to prepare better offers. Based on the study results, LSPs could also concentrate their attention on firms that meet certain criteria, such as a smaller size or a lower degree of internationalization.

Future research

The research topic offers ample possibilities for future research. Based on these study results, highly relevant support possibilities could be investigated deeper by measuring certain constructs, e.g., "personal relationship" separately, for instance, in the Guanxi context through the GRX scale[915]. While the prevalent study was conducted at various regional clusters (asides of China), such as Latin America or Western Societies, a follow-up could be made with an emphasis on certain countries. This is also valid for the industry sector. It is conceivable that results vary for industries where the share of logistics costs for a final product are higher/lower or where the purpose of logistics is primarily final customer distribution instead of production supply. Research methods such as factorial surveys could be employed to address the risk of answering bias and measure the value of the importance of personal relationships. Furthermore, experts in the qualitative study stated that project characteristics, for instance, the volume or the number of involved decision makers, would alter the Guanxi support possibilities during the LSP selection process. As this could not be verified due to insufficient data, a future study could focus on these factors. Finally, considering that values and social norms are in constant change and that it is disputed how Guanxi might evolve, studies should address how the influence of personal relationships on the LSP selection process will likely develop in the future.

---

[913] Shi et al. (2011), p. 501.
[914] Su et al. (2007), pp. 316-317.
[915] Yen et al. (2017), pp. 103-114.

7. Delphi study to forecast the development of the impact of Guanxi on the LSP selection in China

The previous studies provided crucial insights into the current state of Guanxi's impact on the LSP selection process. To also forecast the development of the Guanxi impact, a Delphi study, as described in this chapter, was conducted. As illustrated in Figure 27, this is the final study in this research project.

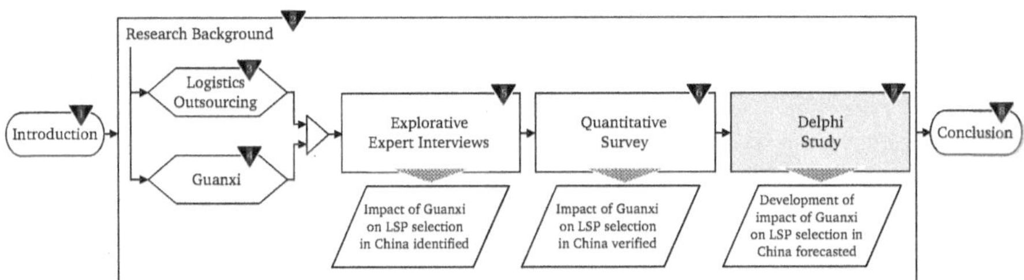

Figure 27: Chapter "Delphi Study" within overall research process
Source: Own figure

In section 7.1, the discussion about the Guanxi influence in the future will be reviewed, and the research questions introduced. Previous studies have not provided conclusive results regarding its future influence, and none was conducted with the focus on LSP selection. To answer the research questions, the Delphi technique was employed (section 7.2). The method is described and evaluated in section 7.2.1. Study participants were selected following the model proposed by Okoli and Pawlowski[916]. This process and the participant demographics are laid out in section 7.2.2. The study follows the method for a ranking type Delphi developed by Schmidt[917]. In section 7.2.3, the corresponding study design and the three related phases are described. The research results in the form of the factors that will impact the future Guanxi influence (7.3.1), the way how the Guanxi influence will develop (7.3.2), and differences among various participant groups (7.3.3) are then stated in section 7.3. They are subsequently discussed in section 7.4. In section 7.4.1, the findings are evaluated and in section 7.4.2 summarized. The final section, 7.4.3, concerns study limitations, implications, and raises possibilities for future research.

7.1. Guanxi influence in the future

There is a general agreement among scholars about Guanxi's importance in China as a crucial factor influencing business. It can, for instance, help to gain access to valuable and relevant information, enhance business, decrease transaction costs, or prevent opportunistic behavior[918]. However, no consensus could be reached regarding a future increasing or decreasing role of Guanxi[919]. One school of thought argues that the influence of Guanxi will decline. With the emergence of a more rational-legal system, the practice of Guanxi will become obsolete[920].

---

[916] Okoli/Pawlowski (2004), pp. 20-23.
[917] Schmidt et al. (2001), pp. 7-13; Schmidt (1997), pp. 763-774.
[918] Shin et al. (2007), pp. 171-172; Luo et al. (2012), pp. 158-161; Standifird/Marshall (2000), p. 30; Wang (2007), p. 85.
[919] Chen et al. (2013), pp. 193-194.
[920] Guthrie (1998), p. 273.

Guanxi is seen as the Chinese form of the personal relations required to accomplish tasks in transforming economies where the inherent traits of socialist economies of shortage have always led to a reliance on personal and particularistic ties instead of rational-legal relationships[921]. Consequently, if the structural and institutional conditions change, Guanxi will not be a necessity anymore. For instance, a study by Xin and Pearce found that private-company executives sought to compensate for their lack of formal institutional support in unstable conditions by cultivating Guanxi with government officials[922] Based on in-depth interviews conducted in Shanghai, Guthrie deduces that the "art of Guanxi practice" will play a diminishing role as the economic transition progresses[923]. His informants stretch that "if you don't follow the procedures and laws, no project will be approved," that "for official procedures, Guanxi is less important today than it ever has been" and that "law is by far the most important thing today[924]." Yao developed an economic model based on which he attributes the issues with Guanxi to the political system that grants and protects privileges[925]. Tan et al. compared companies founded in China's first transitional stage (1979 to 1997) with others founded in the second transitional stage (1997 to 2004). Their findings imply that Guanxi's role arose from a lack of market system trust during China's transition from a centrally planned to a market economy and that it became less important when market system trust based on well-established institutions was established[926]. Comparing Hong Kong and mainland China, Yeung also illustrates the impact of institutional safeguards. While Hong Kong prolonged the rule of law inherited from British governance and thus effectively limited corruption, it is still rampant in mainland China[927]. Investigating China's history, especially during the Han and early Tang Dynasty where the rule of law was common, Yao summarizes that "while culture might affect people's social behavior, a political system together with a legal system of society could introduce a new social norm, changing people's way of thinking and their behavior, and changing culture itself gradually[928]."

A different school of thought argues that the influence of Guanxi will remain or increase in the future. They reason that Guanxi is a deeply rooted part of Chinese culture, existing for thousands of years[929]. For instance, the founder of sociological and anthropological studies in China explains that the fundamental means of control are the institutionalized networks of relationships, "the patriarchal control of the family," "the elders' control of villages," and "the notables' control of other kinds of associational networks[930]." Guanxi is considered a defining element of Chinese culture, "handed down relatively unchanged through time and space[931]." Yang notes that Guanxi as a "repertoire of cultural patterns and resources" is transformed over time, adapting to and shaping new social institutions and structures[932]. She takes Taiwan and post-socialist Russia as examples where Guanxi might have declined in some social domains but flourished in others[933]. Qi takes a different approach by dissecting Guanxi's elements and comparing them to universal cultural concepts to predict the possible future influence. Based

---

[921] Walder (1988), pp. 1-254.
[922] Xin/Pearce (1996), p. 1654.
[923] Guthrie (1998), pp. 281-282.
[924] Guthrie (1998), p. 273.
[925] Yao (2002), p. 293.
[926] Tan et al. (2009), p. 544.
[927] Yeung (2000), pp. 2-8.
[928] Yao (2002), p. 282.
[929] Chen et al. (2013), p. 193.
[930] Fei (1992), p. 29.
[931] Gold et al. (2002), p. 3.
[932] Yang (2002), p. 459.
[933] Yang (2002), p. 459.

on a characterization of Guanxi as a long-term relationship that operates through trust, mutual obligation, and reciprocity, she concludes that Guanxi will pertain despite an intensification of globalization[934]. Tapping also in the notion of globalization, Appelbaum reasons that the influence of Guanxi as part of Chinese business culture will increase together with the growing importance of China in the global economy[935]. Wank carried out field studies among entrepreneurs in Xiamen during China's economic reform and suggested that transformation is an intertwined system of commercial trade and political patronage, where Guanxi consistently plays an important role[936]. Bian and Ang compared how Guanxi networks affect job mobility in Tianjin and Singapore. Although Singapore possesses more mature institutions than mainland China, jobs are similarly channeled through Guanxi networks of exchange relations in both countries[937]. Similarly, Hsing, who analyzed Taiwanese investors' business practices in Guangdong, concluded that the flexibility of the Taiwanese export industry there is founded on firm networks that are based on long-term social relationships among individuals[938].

Together, although various studies have tried to address Guanxi's influence in the future, none provided conclusive results. Also, none were conducted with a focus on LSP selection. To address this gap, the first research question of this study is:

RQd1: How will the influence of Guanxi on the LSP selection in China develop in the future?

Related to the expected influence of Guanxi in the future are factors that will impact it. Some researchers proposed factors such as market maturity, compliance with regulations enforced throughout the recent government's anti-corruption campaign, or cultural adoption[939]. However, previous studies did not identify all factors that might be relevant, especially in the context of LSP selection processes. This is particularly crucial as situational factors are fundamental to understanding the impact of personal relationships and, correspondingly, forecast its future impact[940]. This study set out to address this gap by exploring the following research question:

RQd2: Which factors will impact how the influence of Guanxi on the LSP selection in China will develop in the future?

The studies described in chapters 5 and 6 uncovered that Guanxi's impact on the LSP selection is mitigated by contingency factors such as personal characteristics, firm characteristics, and project characteristics. Thus, likely also factors that will impact how Guanxi's influence on the LSP selection in China will develop in the future is differing across contingency factors. To reveal such differences, setting up groups with the opposing poles of such factors is proposed. For instance, one group could be formed with participants working for SMEs, one with participants from large enterprises and the results compared. However, due to the large number of 10 contingency factors and the exponentially increasing high number of groups, a study would not be manageable anymore, and hence a prioritization must be made[941]. The largest

---

[934] Qi (2013), pp. 308-322.
[935] Appelbaum (1998), p. 171.
[936] Wank (1999), pp. 204-234.
[937] Bian/Ang (1997), pp. 1000-1002.
[938] Hsing (1998), pp. 147-160; Hsing (1998), p. 5.
[939] Guthrie (1998), pp. 281-282; Tan et al. (2009), p. 544; Lin et al. (2016), p. 2; Wederman (2004), pp. 920-921; Tan et al. (2009), p. 544.
[940] Schorsch et al. (2017), pp. 247-254.
[941] For instance, with two opposing factors for sex, "male" and "female," two groups would have to be set up. Adding another characteristic such as age, even with simplified assumptions to create opposing factors, such as "young" and "old," would

differences among groups could be observed between companies from China versus Western Societies, multinational companies versus domestic ones, and LSPs compared to buying firms, cf. section 6.3.2. As this study aims to forecast the development of Guanxi's impact on the LSP selection in China, it is not meaningful to set up groups according to "China" versus "Western Societies." Taken together and under consideration of the contingency factors with the highest group differences in China, research questions RQd3a and RQd3b state:

> *RQd3a: How do factors that will impact how the influence of Guanxi on the LSP selection in China will develop in the future differ among LSPs versus Buying companies?*

> *RQd3b: How do factors that will impact how the influence of Guanxi on the LSP selection in China will develop in the future differ among Multinational versus Domestic companies?*

To answer these research questions, a ranking-type Delphi study was conducted among 57 experts in China.

7.2. Method: Delphi study

One of the most well-known techniques for forecasting is Delphi. In this section, the background of this method will be provided, and the technique evaluated. Afterwards, the participant demographic and the used selection process will be introduced. The last section will then outline the study procedure.

7.2.1. Background and evaluation

Delphi is a method to effectively structure a group communication process so that a group of individuals can deal with a complex problem[942]. It is particularly suitable for cases where expert opinion is the best evidence available, especially in forecasting and theory building[943]. Delphi is therefore typically applied to "complex and under-explored phenomena" in science and technology forecasting as well as business forecasting[944]. However, as it can be seen as a communication process, the method can be applied to a broad field of endeavors[945]. Previously, it was used to derive standards for information needs, identify risks, analyze cost-effectiveness, identify drivers and barriers of good practices implementation, or develop causal relationships[946]. Due to its versatility and flexibility regarding design and location, the method is also helpful at various stages in a research process[947]. For instance, to identify a research topic, specify research questions, select variables of interest or preliminary identify causal relationships[948]. As information in a Delphi study is obtained from experts with relevant experience, the likelihood that the resulting theory will hold across multiple contexts and settings is increased[949].

---

mean that four groups "male young," "male old," "female young" and "female old" have to be set up. With another characteristic this would then become eight groups etc.

[942] Linstone/Turoff (1975a), p. 3.

[943] Martino (1993), pp. 22-125; Lummus et al. (2005), p. 2688; Kudłak et al. (2018), pp. 282-285.

[944] Kudłak et al. (2018), pp. 282-285.

[945] Linstone/Turoff (1975a), pp. 4-5.

[946] Elmer et al. (2010), p. 107; Schmidt et al. (2001), p. 6; Quade (1967), pp. 1-26; Giunipero et al. (2012), p. 258; Palacios Marqués/José Garrigós Simón (2006), p. 143.

[947] Keeney (2010), pp. 231-232; Okoli/Pawlowski (2004), p. 27.

[948] Okoli/Pawlowski (2004), p. 27.

[949] Okoli/Pawlowski (2004), p. 27.

Linestone and Turoff propose that one or more of the following properties lead to the need for employing Delphi[950]. The problem does not lend itself to precise analytical techniques but can benefit from subjective judgments on a collective basis. The individuals needed to contribute to examining a broad or complex problem have no history of adequate communication and may represent diverse backgrounds in their experience or expertise. More individuals are needed than can effectively interact in a face-to-face exchange. Time and cost make frequent group meetings infeasible. A supplemental group communication process can increase the efficiency of face-to-face meetings. Disagreements among individuals are so severe or politically unpalatable that the communication process must be refereed and/or anonymity assured, or the participants' heterogeneity must be preserved to assure the validity of the results, i.e., avoiding domination by quantity or strength of a particular personality[951].

Throughout its history, the method evolved into several *variations* with different aims, target participants, and administration styles, so that it is more accurate to speak of Delphi techniques in the plural instead of the singular form[952]. For example, a "Policy Delphi" aims to generate opposing views on policy and potential resolutions, where policymakers are selected to obtain diverse opinions. The administration can adopt several formats, including a meeting of participants as a group. In a "Real-time Delphi," opinion should be elicited, and consensus gained in real-time from selected experts. This form can be administered through computer technology that participants use in the same room[953]. The "Ranking-type Delphi" design, which is also used in this study as it is most suitable to answer the posed research questions, is commonly employed to achieve group consensus about the relative importance of issues. Classically, it consists of a process of identification and elaboration of concepts, followed by a classification of them[954]. All these variations share certain characteristics[955]. It is a repetitive process, where experts are asked the same question at least twice, so they can re-consider their answers and take the feedback from other experts into consideration[956]. The anonymity of participants and their answers is maintained to minimize influences caused by the experts' personality or status[957]. All information exchange between experts is controlled so that irrelevant information can be filtered by the study coordinator[958]. Feedback is provided as a group statistical response, where the final answer is always a computed outcome of the individual answers[959]. Reaching consensus among the experts used to be compulsory for all types of Delphi studies, but is now not considered mandatory anymore[960].

Since the first study was performed according to the Delphi methodology in the 1950s, the technique developed through various phases. In the 1960s, it was in a stage of novelty, where a wide variety of studies employing it was published. Then in the 1970s and 80s, the method entered a scrutiny stage, where its validity and reliability were questioned[961]. For the past

---

[950] Linstone/Turoff (1975a), pp. 4-5.
[951] Linstone/Turoff (1975a), pp. 4-5.
[952] Rowe/Wright (2011), p. 1487.
[953] Keeney (2010), pp. 231-232; Hasson/Keeney (2011), p. 1697.
[954] Okoli/Pawlowski (2004), p. 16.
[955] Landeta (2006), pp. 468-469.
[956] Hasson/Keeney (2011), p. 1696.
[957] Hasson/Keeney (2011), p. 1696.
[958] Hasson/Keeney (2011), p. 1696.
[959] Hasson/Keeney (2011), p. 1696.
[960] Hasson/Keeney (2011), p. 1696; Landeta (2006), pp. 468-469.
[961] Quade (1967), pp. 1-26; Dalkey/Helmer (1963), p. 458; Rieger (1986), pp. 195-197; Winkler/Moser (2016), pp. 63-76.

decades, Delphi has been in a stage of continuity with a remaining application frequency, as researchers provided guidelines on how the technique's use can be enhanced[962].

The Delphi method has a variety of *advantages* compared to single opinions. The full information available to a group is greater than to an individual, and adding members does not inherently destroy information open to a group. The group can consider more factors than an individual can, and a group is less susceptible to bias caused by the pure repetition of arguments[963]. Moreover, with Delphi, a group is less vulnerable to the influence of dominant individuals and members who concentrate on winning arguments instead of reaching the best possible group output[964]. The method also offers several advantages compared to other research methods based on direct group interaction. It reduces the influence of some undesirable psychological effects among the participants, such as inhibition, and researchers receive selective feedback of relevant information[965]. There are more extensive consideration thanks to the repetition loops; results are statistical, the methodology is flexible, and the execution simple[966]. Some key differences to traditional surveys are the required sample size for statistical power to obtain significant findings, the nature of individual compared to group responses, and the lower risk for non-response issues[967]. Nevertheless, methodical *challenges* are attributed to the Delphi method concerning the group of predictors, the questions asked, and the procedure itself[968]. One of the most frequently posed challenges is the definition of what constitutes an expert, as well as the selection and identification of them[969]. Some scholars also question if consensus is the right way to approach the truth, and if assessing the accuracy and reliability of the method is straightforward[970]. During the group process itself, experts who act irresponsibly are conferred impunity, while constructive-minded ones do not receive reinforcement and motivation through the support and social approval of the other group members[971]. Interaction is allowed only in controlled form through written feedback, and possible interrelations between forecast incidents are not considered[972]. Asides from these methodological weaknesses, inadequate research results are also attributed to a faulty application of the technique, for instance, due to a lack of explanation, poorly formulated questions, or insufficiently analyzed results[973].

7.2.2. Participants

This section will describe the selection process and display the participants' demographics.

Selection process

Selecting qualified experts is considered a critical success factor for conducting Delphi studies[974]. Experts can be defined in numerous ways based on their position in a hierarchy,

---

[962] Önkal et al. (2017), pp. 280-297; Meijering/Tobi (2016), pp. 166-173; Rowe/Wright (2011), pp. 1489-1490; Paré et al. (2013), pp. 210-216.
[963] Martino (1993), pp. 22-125; Lummus et al. (2005), p. 2689.
[964] Martino (1993), pp. 22-125; Lummus et al. (2005), p. 2689.
[965] Landeta (2006), p. 469.
[966] Landeta (2006), p. 469.
[967] Okoli/Pawlowski (2004), pp. 19-20.
[968] Kaplan et al. (1950), p. 108.
[969] Sackman (1974), pp. 33-42; Linstone/Turoff (1975a), pp. 3-7.
[970] Martino (1970), pp. 63-64; Weaver (1972), pp. 1-42; Sackman (1974), pp. 33-42; Dalkey/Helmer (1963), pp. 458-466.
[971] Ven, Andrew H Van De/Delbecq (1974), pp. 616-620; Becker/Bakal (1970), pp. 208-209.
[972] Linstone (1975), pp. 559-571; Milkovich et al. (1972), pp. 382-388.
[973] Landeta (2006), pp. 468-469; Rieger (1986), pp. 197-202.
[974] Sackman (1974), pp. 33-42; Linstone/Turoff (1975b), pp. 3-7.

particular experience, public acknowledgment, i.e., they have published or lectured comprehensively on a specific topic, or as they are considered experts by the group of people to be investigated[975]. Asides from possessing knowledge about the research area, participants in a Delphi study should have the motivation to engage with the inquiry process and be able to articulate judgments[976]. To reach a consensus faster or compare different opinions, experts with different perspectives should be divided into different groups called "panels[977]."

Okoli and Pawlowski proposed a five-step model to select study participants, which has its origins in the work of Delbecq et al. and is a rigorous procedure to ensure the identification of relevant experts[978]. This model was also employed in this study. Step 1, *prepare knowledge resource nomination worksheet (KRNW)*. The purpose of the KRNW is to help to categorize experts before they are identified so that no vital class is overlooked. Following the results of the previous studies and the research questions, the firm characteristics "type" and "internationalization degree" are most relevant for this study. The classes were therefore split accordingly into "LSPs," "Buying companies," "Multinational companies," and "Domestic companies," following the definitions set out in section 6.1. This leads to the four combinations a) Buying company, domestic (BD), b) Buying company, multinational (BM), c) LSP, domestic (LD), and d) LSP, multinational (LM). Step 2, *populate KRNW with names*. Names of potential experts, according to the previously defined classes, should be entered into the worksheet. For each category, contacts from the chair of Global Supply Chain Management at Tongji University and the researcher's personal network were used. Although it was possible to find experts for all panels, the goal of identifying at least 15 per panel was not reached[979]. Step 3, *first-round contacts-nominations for additional experts*. Identified experts should be contacted and asked to provide basic information and to nominate further experts. The already identified possible panelists received a brief description of the research project, and that they were identified as experts on the impact of personal relations on the LPS selection in China. They were asked to provide some demographic information, rate their level of expertise in the subject matter, and nominate other suitable experts. With this step, it was possible to identify a sufficient number of qualified experts for every panel[980]. Step 4, *ranking experts by qualifications*. The various experts' lists should be ranked according to their qualifications to prioritize them for an invitation to the study. Correspondingly, the names on the KRNW were ranked within the four categories[981]. Step 5, *inviting experts to the study*. Panelists should be contacted, and the subject of the research project as well as the procedures required for it explained. Only experts with at least five years of experience and medium-high expertise about the LSP selection process were considered for this study. They were invited into separate groups in WeChat, a widely used social messaging application in China, following the four categories[982].

Panelists

In total, 57 experts participated in the study that lasted from 2018-01-11 to 2018-02-17, where two dropped out throughout the iterations. All panelists had at least five years of work experience, and most held managing positions in their companies. Due to the research topic's

---

[975] Baker et al. (2006), pp. 61-62; Mead/Moseley (2001), pp. 10-11.

[976] Day/Bobeva (2005), p. 108.

[977] Okoli/Pawlowski (2004), pp. 19-20.

[978] Okoli/Pawlowski (2004), pp. 20-23; Delbecq et al. (1975), pp. 87-89.

[979] Okoli/Pawlowski (2004), pp. 20-23.

[980] Okoli/Pawlowski (2004), pp. 20-23.

[981] Okoli/Pawlowski (2004), pp. 20-23.

[982] Okoli/Pawlowski (2004), pp. 20-23.

sensitivity, anonymity was promised, so that no individual names or names of the employers are shared. The demographics of the panelists are shown in Table 21 and subsequently discussed.

Table 21: Delphi panel demographics

Panel	BD	BM	LD	LM	Total
Participants					
Originally identified	4	19	6	11	40
After nomination of additional experts	16	22	24	15	77
After expertise evaluation	12	14	19	12	57
Job title					
General Manager			2		2
Director	1	1	2		4
Manager	11	10	15	11	47
Others		3		1	4
Job functions					
Purchasing	2	5			7
Operations	2	2	2	1	7
Supply chain management	2	1			3
Logistics	6	6	3	5	20
Sales			8	6	14
Other			6		6
Work experience					
5-6	2	3	3	1	14
7-8	5	4	1	1	12
9-10	1		1	3	7
11-12	1	2	2	2	9
13-14		2	5	2	9
15-16	1	1	1	3	7
17-18			2		3
19-20	2	1	2		5
>20		1	2		3
Level of expertise					
High	8	9	12	8	44
Medium	4	5	7	4	25

BD. Four experts were initially identified for this panel. After the nomination of additional experts, a total of 16 experts could be found. Following the evaluation, 12 were invited into the WeChat group chat to participate in the study. All experts partook throughout the course of the whole Delphi study. Most were working on Manager level in logistics functions and had 7-8 years of work experience. The experts were also asked to self-assess their experience on a scale from 1 to 4, with 1 meaning "no expertise regarding the selection process of LSPs" and 4 meaning "high expertise." All 12 experts had either medium or high expertise about the selection process of LSPs.

BM. 19 experts were initially identified for this panel, and in step 3 of the selection process, three more added. After the evaluation, 14 were invited to participate in the study. Two experts remained only in the first and second rounds, while 12 participated from the first round until the end of the study. Most participants in this group were on Manager level, working in logistics and purchasing functions. They often possessed 7-8 or 11-14 years of work experience. The self-evaluation indicated that nine of them had high and five medium expertise about the topic.

LD. In phase 2 of the selection process, six experts could be identified. After the nomination of additional experts, 18 more were identified. Following the evaluation of these in total 24 potential participants, 19 were invited to join. All 19 selected experts participated from the first until the last round of the Delphi study. Most participants worked on Manager level, two on Director and two on General Manager level. Sales was the most common job function, and experts possessed a variety of years of work experience. Most experts had a high level of expertise in the selection process of LSPs.

LM. 11 experts were identified in phase 2 of the selection process. These experts then identified three additional ones. In total, 12 had a suitable profile for the LM panel and were invited to participate in the study. All 12 participated from the first until the last round of the study. Most participants had a manager job title and worked in a sales or logistics function. The experts in this panel possessed mostly more than eight years of work experience and a high level of expertise about the topic.

7.2.3. Procedure

The study approach follows the procedure for a ranking type Delphi developed by Schmidt[983]. As recommended by Paré et al., it was pretested with a panel of student volunteers[984]. This section will describe the study design and the different phases.

Study design

Recommendations about the ideal *number of rounds* that should be conducted in a Delphi study range from two up to 10[985]. While generally, the accuracy of results increases with the number of rounds, there is also an increased risk that experts drop out[986]. Previous studies showed that the most critical improvements are made between the first and second iteration[987]. A common way to decide when to stop iterations is when either *consensus* among the experts is reached or after two consecutive rounds without significant change[988]. Here, consensus can be measured with "Kendall's coefficient of concordance," also known as "Kendall's W[989]." A Kendall's W value

---

[983] Schmidt et al. (2001), pp. 7-13; Schmidt (1997), pp. 763-774.

[984] Paré et al. (2013), p. 215.

[985] Goluchowicz/Blind (2011), p. 1530.

[986] Goluchowicz/Blind (2011), pp. 1538-1539.

[987] Woudenberg (1991), p. 140.

[988] Schmidt (1997), pp. 770-771.

[989] Schmidt (1997), p. 765; Kendall's W is calculated as follows (Kendall/Smith (1939), pp. 275-287). The factor i is given the rank $r_{i,j}$ by expert number j; in total there are n factors and m experts.

In this case the total rank given to factor i is:

$$R_i = \sum_{j=1}^{m} r_{ij}$$

The mean value of the ranks, R_{Mean}, is:

of typically W > 0.7 indicates strong agreement[990]. If Kendall's W reaches a value of W > 0.5 and there is no significant change after two consecutive rounds, moderate agreement in the panel can be assumed[991].

To benefit from the Delphi technique's advantages such as reduced pressure for conformity and reduced influence of dominant individuals, participating experts should only alternate their opinion based on new information[992]. Therefore, it is crucial to provide the right type of *feedback* after each iteration. Different feedback types are possible, ranging from purely statistical summaries of the results to detailed descriptions of other experts' feedback and reasoning. Previous studies showed that panelists who receive reason-based feedback are less likely to change their opinions in the following rounds, whereas statistical feedback leads to more changes and conformity[993]. In a ranking-type Delphi study, the ideas are generated in the first round through brainstorming, which need to be categorized and summarized before shown to the experts in the following rounds. To ensure the validity of this feedback, a content analysis should be conducted, followed by verification from the experts[994]. In international studies, it also leads to better results to provide questions and feedback translated[995].

Possible challenges with Delphi studies include the long response time and the attrition of participants. Some specialized *online tools* can support mitigating these issues, but accessing them in China is often not feasible due to local internet security policies[996]. In addition, contextual factors such as social and cultural norms pose a further challenge for research conducted in China[997]. To address these obstacles, a novel approach using the social communication platform WeChat together with the online survey tool WJX was employed[998]. WeChat is a widely used mobile phone application in China, with over 80% penetration rate among internet users[999]. It provides a wide range of functionalities, including free text/voice messaging services, video chats, photo sharing, and location information exchange[1000]. The function that was mainly used for this research project was "group chats," where people can have a chat conversation and send multimedia messages to a defined group of people[1001]. Group chats were used to distribute questionnaires, feedback, and reminders. The survey tool WJX

$$R_{Mean} = \frac{1}{n} \sum_{i=1}^{n} R_i$$

Then the sum of squared deviations, S, is:

$$S = \sum_{i=1}^{n} (R_i - R_{Mean})^2$$

With those definitions, Kendall's W is calculated as follows:

$$W = \frac{12\,S}{m^2(n^3 - n)}$$

---

[990] Marozzi (2014), pp. 1843-1844.
[991] Schmidt et al. (2001), p. 13.
[992] Rowe et al. (2005), pp. 383-384; Woudenberg (1991), p. 140.
[993] Rowe et al. (2005), p. 382; Woudenberg (1991), p. 140.
[994] Kembro et al. (2017), p. 80; Schmidt (1997), p. 769.
[995] Frewer et al. (2011), p. 1522.
[996] Deibert (2002), pp. 143-159.
[997] Mei/Brown (2017), pp. 721-722.
[998] Gan/Wang (2015), pp. 351-363; Mei/Brown (2017), pp. 722-730.
[999] Mei/Brown (2017), pp. 722-730.
[1000] Xu et al. (2015), p. 21; Su (2016), pp. 233-234.
[1001] Qiu et al. (2016), pp. 311-320; Wu/Wan (2014), p. 5.

was employed to create the questionnaire and collect answers[1002]. It provides the required functionality for conducting the various phases of a ranking-type Delphi study, including open-ended questions, multiple-choice questions, and a ranking function. WJX also provides a sharing functionality to WeChat group chats, where survey links can be disseminated[1003]. Using this innovative approach maximized convenience for participants in China so that the time frame to conduct the study could be shortened and the dropout rate compared to typical Delphi studies reduced[1004]. All used questionnaires are shown in Appendix d1.

Phases

According to Schmidt, the ranking-type Delphi design is conducted in three phases, which are *Brainstorming, Narrowing down,* and *Ranking*[1005].

<u>1. Brainstorming</u>

In the first phase, all participants are asked to submit as many factors as possible to maximize the chance that all critical aspects are covered. In this stage, it is crucial to ask the participants to describe each factor so that similar factors mentioned by several experts with different terms (or in a different language) can be identified[1006].

Then, the mentioned factors must be categorized, and answers consolidated into a single list, which is to be verified by the experts to ensure that all ideas are represented. If different panels are participating in the study, this first step must be done by all groups collectively, as it otherwise would be challenging to compare results for different panels. The first step should be repeated until all participants have approved the list[1007].

In the prevalent study, the target of phase 1 was to collect all factors that are possibly impacting how the influence of Guanxi on the LSP selection in China will develop in the future. Therefore, first, to ensure that every expert had the same understanding of the research process and its underlying assumptions, a short introduction including the used definitions, previous study findings, and the process of the ranking type Delphi method was distributed to the selected participants. Second, a brief questionnaire was sent to the experts to validate their demographic information and ensure that they were all categorized in the right panels. Third, for the actual brainstorming, experts were asked to list as many factors as possible and briefly explain their reasoning. This step had to be completed jointly by all panels before the content analysis, and the next round could begin. All experts submitted their answers within two weeks, and 156 factors could be collected, which are listed in Appendix d2. Fourth, a content analysis was conducted where duplicate answers were eliminated, and similar ones were categorized. For this, answers provided in Chinese were translated into English by Chinese native speakers with excellent command of the English language. The content analysis was then conducted individually by two researchers. After the analysis was done individually, the results were carefully compared, differences among the given answers identified, and subsequently discussed until consensus could be reached. This led to a reduction of the factors from 156 to 31. The 31 factors were grouped into factors that predict a reduction, increase, or stable remaining of Guanxi's future impact on the LSP selection process and afterwards sent back to

---

[1002] Mei/Brown (2017), pp. 722-730.

[1003] Mei/Brown (2017), pp. 721-722.

[1004] Huckfeldt/Judd (1974), pp. 75-88. This excludes "real time Delphi" approaches which are conducted in real time, cf. Gordon/Pease (2006), pp. 321-333; Gnatzy et al. (2011), pp. 1681-1694.

[1005] Schmidt (1997), pp. 763-774; Schmidt et al. (2001), pp. 7-13.

[1006] Häder (2014), pp. 153-163.

[1007] Schmidt (1997), pp. 768-769.

the panels for verification. The participants approved all identified factors and their correct categorization. The results are listed in Appendix d3 and d4. To ensure generalizability and transferability, the findings were finally compared to the relevant literature and the previous studies' results[1008]. They are shown in section 7.3, "Results," and section 7.4, "Discussion."

<u>2. Narrowing down</u>

In the second phase, the list created during the first phase must be reduced to a "manageable size." Out of the usually large number of identified issues, a reduced list can be created, making it easier to rank. Schmidt suggests sending the previously consolidated list to the participants in random order and to ask them to choose the most important factors[1009]. All factors not selected as important by at least 50% of the participants can then be eliminated. The list should typically contain less than 20 items that the experts could agree on[1010]. If different panels are participating, the panels must be separated at the beginning of the second phase. Since the consolidated list from the first phase was developed by all panels together, it is possible to compare the various panels' results[1011]. In the prevalent study, experts were asked to choose factors from the list they consider most important, with the hint that a maximum of 10 should be selected. Replies were submitted within one week, and factors were eliminated, not selected by at least 50% of the experts in a panel.

In the BD panel, the nine factors 1, 2, 3, 4, 9, 12, 20, 21, and 31, were chosen by at least 50% of the experts. Figure 28 shows which factors were selected how often by the experts in this panel.

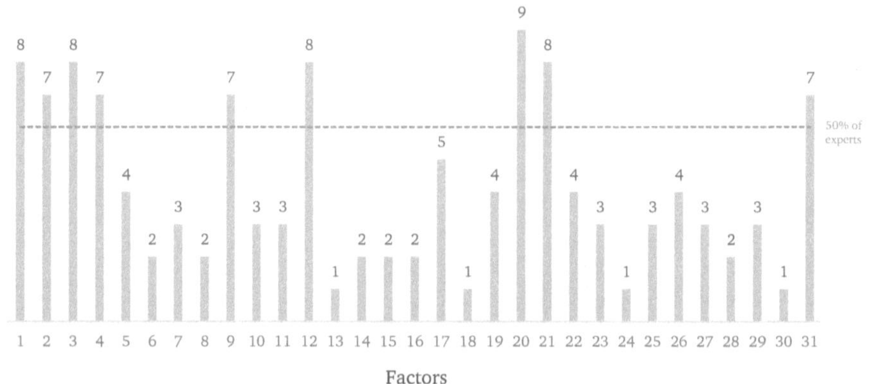

Figure 28: Factors selected by BD panel
Source: Own figure

The nine factors 1, 2, 3, 4, 9, 18, 20, 27 and 31, were chosen by at least 50% of the experts in the BM panel. Figure 29 shows which factors were selected how often by the experts.

[1008] Hasson/Keeney (2011), pp. 1695-1704.
[1009] Schmidt (1997), pp. 769-770.
[1010] If there are many more items, it is recommended to repeat the second phase also for the shortened list until the list is successfully reduced to a "manageable size," Schmidt (1997), pp. 769-770.
[1011] Schmidt (1997), pp. 769-770.

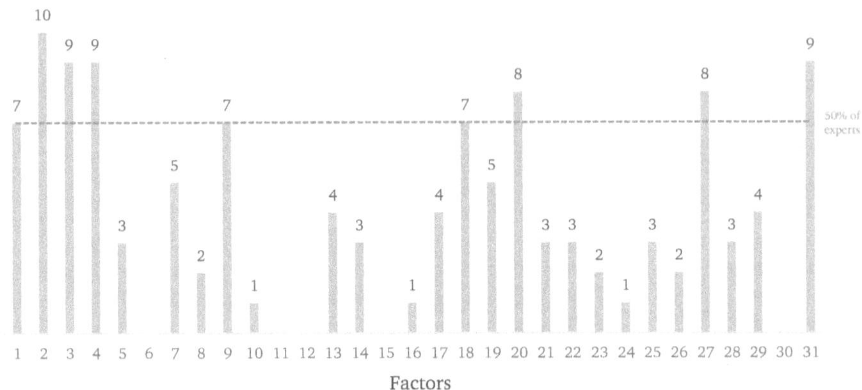

Figure 29: Factors selected by BM panel
Source: Own figure

In the LD panel, the ten factors 4, 9, 17, 19, 20, 21, 23, 26, 29, and 31, were chosen by at least 50% of the experts. Figure 30 shows which factors were selected how often by the experts.

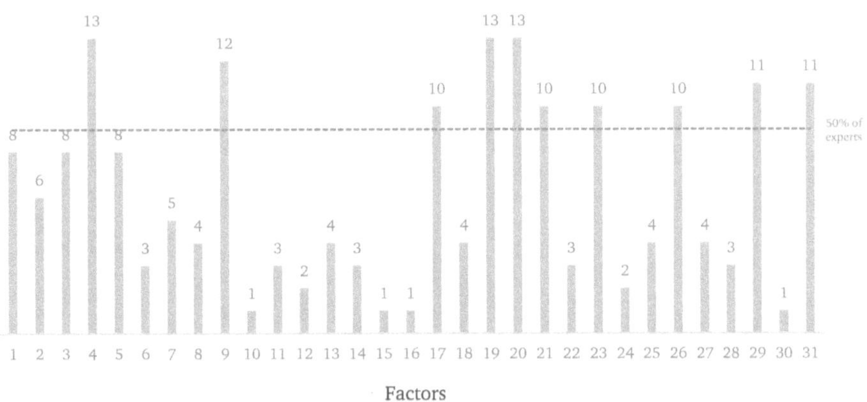

Figure 30: Factors selected by LD panel
Source: Own figure

In the LM panel, the nine factors 1, 2, 9, 17, 18, 20, 21, 23, and 31, were chosen by at least 50% of the experts. Figure 31 shows which factors were selected how often by the experts.

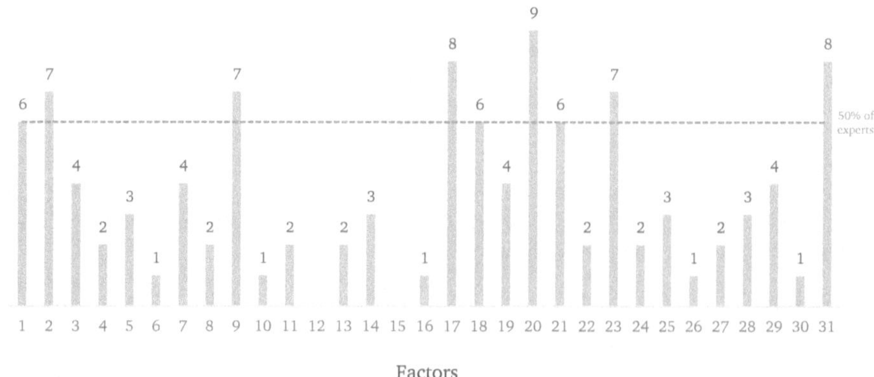

Figure 31: Factors selected by LM panel
Source: Own figure

3. Ranking

In the third phase, the experts are asked to rank the shortened list, and the ranking results are used to create a consensual ranking. A standard method to create this ranking is Kendall's W, which is simple in its application and easy to understand. In the first round of the ranking phase, the items on the list are provided in a random order to the experts. In the following rounds, feedback is given by listing the items in the order of their average ranks within each panel. Experts are asked to re-rank the items on the list until W as an indicator for the degree of consensus reaches a specific value, or when the researcher decides that an additional round would not lead to an improvement that would justify the additional effort[1012].

In this study, every panel was asked to rank the previously selected factors from a randomized list. After each round, Kendall's W was calculated and assessed if either a value equal to or above 0.7, indicating strong consensus, or a value equal to or above 0.5 for two consecutive rounds, indicating a lack of progress, was reached. In the first iteration, neither of these criteria could be reached, so feedback from the first iteration was provided and distributed to each panel. This led to two experts from the BM panel dropping out. The mean ranking order for each panel was sent back as feedback, and the experts were asked to rethink their answers considering this new input. For the second iteration, the factors were presented in the order of the mean ranking for each panel as a reference. After the second iteration, all panels achieved a W value higher than 0.7. Hence, no further iteration was needed, and the phase could be concluded.

In the *BD panel*, the level of agreement in the first iteration was medium to low (W = 0.327), while in the second iteration, it improved to a high level (W = 0.731). The most significant changes from the first to the second iteration include factor 3, which dropped from rank 5 to 9, factor 21, which increased from rank 9 to 7, and factor 1, which increased from 8 to 6. Also, factors 4 and 12 increased one rank each, from 4 to 3 and 6 to 5. Factor 9 and 31 both decreased one rank from 3 to 4 and 7 to 8. A moderate consensus was reached in the *BM panel* after the

---

first iteration (W = 0.528) and improved to a high consensus after the second (W = 0.743). Factor 3 and 18 dropped one rank each from 8 to 9 and 4 to 5, respectively. Factor 31 increased from rank 9 to 8 and factor 20 from rank 5 to 4. The *LD panel* initially had a low level of agreement after the first round (W = 0.269) but reached consensus after the second (W = 0.760). Factor 23 decreased one rank from 3 to 4, and factor 20 increased from 4 to 3. In the *LM panel*, initially, a moderate to high level of consensus could be reached after the first round (W = 0.613) and, subsequently, a high level in the second (W = 0.714). Changes include a decrease of factor 1 from rank 5 to 7 and an increase of factor 23 from 8 to 5. Factor 9 declined from rank 7 to 8 and factor 20 from 3 to 4, while factor 21 increased from 4 to 3. All Kendall's W calculations are shown in Appendix d5.

7.3. Results

This study's target was to identify which factors will impact how Guanxi's influence on the LSP selection in China will develop in the future and how. In this section, the factors that will impact the future Guanxi influence, how the Guanxi influence will develop, and the differences among various panels will be listed.

7.3.1. Factors impacting the future Guanxi influence

The list of factors identified in phase 1 is shown in Table 22. It is categorized into a set of three groups, based on an expected reduced, increased, or remaining influence of Guanxi on the LSP selection in China.

Table 22: Factors impacting the future Guanxi influence

Impact of Guanxi influence	Factor
Reduced	1. Increased usage of artificial intelligence will lead to less influence of humans in a tender process. Guanxi is between humans.
	2. Compliance will become more and more important in corporations. Guanxi might be seen as not compliant.
	3. Enforced laws and regulations against corruption will increase and Guanxi might be related to corruption.
	4. Digitalization will make the selection processes more transparent and fair. Guanxi is mainly operating in grey areas and has unfair impacts.
	5. The Chinese logistics market will become more mature and Guanxi helps to compensate market inefficiencies.
	6. Women see Guanxi as less important part of business and life compared to men. Women will increasingly take over decision power.
	7. Purchasing departments will become more and more centralized and take over decision power. This makes relationship building very difficult.

8. Profitability will become increasingly important in the future and establishing and maintaining Guanxi is expensive.
9. Purchasing processes will become more and more standardized and limit possibilities to take influence through Guanxi.
10. Efficiency will become increasingly important in the future and establishing and maintaining Guanxi is time intensive.
11. Younger generations (especially post-90) see Guanxi as less important part of business and life compared to previous generations. These younger generations will increasingly take over decision power.
12. Logistics will become a more important function in a company in the future. In functions with high focus Guanxi impact is actively reduced.
13. Cooperation and communication becomes less dependent on relationships between individuals. Guanxi is related to the relationship between individuals.
14. Internationalization of the logistics market will lead to an adoption of international business practices.
15. Managers see Guanxi as less important part of business and life compared to before.
16. Associate's turnover will increase in the future. It will be therefore more difficult to build up Guanxi relationships.
17. The reputation of a corporation becomes increasingly important in the future. Using practices related to corruption harms a corporation's reputation when discovered. Using Guanxi is sometimes connected with corruption.

Increased	
	18. The business environment is becoming more volatile, uncertain, complex and ambiguous. Guanxi helps to reduce risks in such a business environment.

18. The business environment is becoming more volatile, uncertain, complex and ambiguous. Guanxi helps to reduce risks in such a business environment.
19. Contractees will use more strategic purchasing in the future. Characteristic for strategic purchasing is a closer relationship with suppliers and Guanxi is about building relationships.
20. Services will become more customized in the future and mutual understanding through information exchange is the base for service customization. Guanxi supports information sharing between business partners.
21. Trust is the basic foundation of business and will become more important. Guanxi supports building up trust.

	22. Profitability will become increasingly important in the future and Guanxi can lead to competitive advantages.
	23. Service quality will become increasingly important in the future. An important element of Guanxi is the exchange of favors beyond contractual agreements which will improve the service quality.
	24. Selection processes will become increasingly more efficient in the future. Guanxi leads to a more efficient information exchange.
	25. Logistics outsourcing will become more important in the future as it enables companies to concentrate on core competencies. Guanxi is about relationships and supports outsourcing.
	26. Contractor's reputation will become a more important decision criterion in the future. Companies with high Guanxi have a higher reputation.
	27. Market shares will become increasingly important in the future. Guanxi leads to willingness to invest into the growth of partners.
	28. The profitability pressure will be higher in the future. This might lead to a market consolidation including the bankruptcy of contractors. To survive as contractor a strong relationship with the contractee is important and Guanxi is about relationships.
Remaining	29. Guanxi will always be an essential part of the Chinese culture.
	30. No future factor will affect the impact of Guanxi on the selection process.
	31. Personal feelings will always remain in business and Guanxi is related to personal feelings.

This list addresses the second research question, which factors will impact the development of the future influence of Guanxi on the LSP selection in China. Out of these factors, 17 predict a reduction, 11 an increase, and three a stable remaining future impact of Guanxi on the LSP selection process in China.

In phase three of the Delphi study, each panel had to rank factors according to their relative importance. Table 23 provides an overview of the ranking results per panel.

Table 23: Delphi factor ranking

Rank	BD	BM	LD	LM
1	2	4	19	2

2	20	9	21	17
3	4	1	20	21
4	9	20	23	20
5	12	18	9	23
6	1	2	4	18
7	21	27	29	1
8	31	31	26	9
9	3	3	17	31
10			31	

The *BD panel* ranked the factors 2, 20, 4, 9, 12, 1, 21, 31, and 3 from highest to lowest. Dominant themes were first, an increase in compliance and enforced laws and regulations against corruption with related factors being selected on rank 1 and 9, and second, an increase in digitalization of selection processes and an increased usage of artificial intelligence. In addition, further standardization of purchasing processes and the increasing importance of the logistics function are forecasted. All these factors will lead to a decreased influence of Guanxi. Another chosen factor was the prediction that services will become more customized in the future, while mutual understanding through information exchange is the base for service customization. Guanxi supports this needed information sharing between business partners, and hence its influence will likely increase. The panel also selected the factor that highlights trust as a basic foundation of business and Guanxi's positive impact on building it up, leading to an increased Guanxi influence. Finally, the panel acknowledged that personal feelings will always remain in business, and as Guanxi is related to personal feelings, it will maintain its influence. The *BM panel* ranked the factors 4, 9, 1, 20, 18, 2, 27, 31, and 3 from highest to lowest. For this panel, an increase in the digitalization of selection processes and an increased usage of artificial intelligence is most important, with the linked factors selected on rank 1 and 3. The experts also predicted a further standardization of purchasing processes, limiting possibilities to take influence through Guanxi. However, the panel also chose factors that predict an increase in the importance of Guanxi. These are the positive impact of Guanxi on information sharing among business partners to enable the future increasing service customization, the risk reduction through Guanxi in a VUCA environment as well as the Guanxi-supported willingness to invest in the growth of partners. The panelists additionally selected factors that indicate a rise in corporations' compliance regulations and an increase of enforcement of laws and regulations against corruption, although at the lower ranks 6 and 9. It was further acknowledged that personal feelings will always remain in business, and as Guanxi is related to personal feelings, it will maintain its influence. The *LD panel* ranked the factors 19, 21, 20, 23, 9, 4, 29, 26, 17, and 31 from highest to lowest. This panel selected dominantly factors that predict an increase of Guanxi's importance, mainly due to the increasing importance of service customization and quality, as well as strategic purchasing. Guanxi facilitates a closer relationship with suppliers, supports better information sharing between business partners, and the exchange of favors beyond contractual agreements. The experts also highlighted on rank two that trust as a basic foundation of business is supported by Guanxi and consider Guanxi as well as personal feelings an essential part of Chinese culture and business. Moreover, they predicted that purchasing processes will become more standardized and digitalized, thus limiting the possibilities of Guanxi influence. The panel also selected two factors related to the increasing importance of a company's reputation, although being ambiguous about the possible impact on the Guanxi influence. On the one side, they perceive that companies with Guanxi have a higher reputation; on the other side that using Guanxi practices related to corruption might be harmful. The *LM panel* ranked the factors 2, 17, 21, 20, 23, 18, 1, 9, and 31 from highest to lowest. The experts saw an increase in compliance activities in corporations as key factor, ranked as first. They connected it with the increasing importance of a corporation's reputation and the risk of its loss, selecting the associated factor on rank 2. However, they also predicted that with trust being the foundation of business and the increasingly volatile, uncertain, complex, and ambiguous business environment, the Guanxi impact might increase. This is also connected with the selected factors regarding an increase in service customization and service quality, where Guanxi can support information sharing and favor exchanges beyond contractual agreements. The experts further recognized that a more standardized purchasing process and increasing usage of artificial intelligence in tender processes will reduce the possibility of taking influence. Lastly, they acknowledge that personal feelings will always remain in business and that Guanxi's influence will, therefore, somewhat remain.

7.3.2. Future development of Guanxi influence

RQd1 asks how the influence of Guanxi on the LSP selection in China will develop in the future. Various indicators can be used to provide an answer.

Number of factors identified in brainstorming session. In phase one of the Delphi study, all panels were jointly asked to state factors that will impact how Guanxi's influence on the LSP selection in China will develop in the future. The provided factors were then categorized based on whether the Guanxi impact will be reduced, increased, or remain the same. In the first category, 17 factors could be identified, while in the second category 11, and in the third category three. Therefore, the number of provided factors in the brainstorming session indicates that Guanxi's impact will be reduced in the future.

Number of factors selected in narrowing down phase. In phase two of the Delphi study, the individual panels had to choose factors that they consider as most important. The results are shown in Table 24. In the category Guanxi impact will be reduced, the BD panel chose 6 factors, BM 5, LD 3, and LM 4. In the category Guanxi impact will be increased, BD selected 2 factors, BM 3, LD 5, and LM 4. Moreover, in the category Guanxi impact will remain the same, the BD panel selected 1 factor, BM 1, LD 2, and LM 1. In total, 18 times factors in the category reduced were selected, 14 times factors in the category increased, and 5 times in the category remain. Hence, based on this indicator, the Guanxi impact will be reduced in the future.

Table 24: Number of selected factors per category

Panel												Reduced (Rd)	Increased (In)	Remain (Re)
BD	Factor	02	20	04	09	12	01	21	31	03				
	Category	Rd	In	Rd	Rd	Rd	Rd	In	Re	Rd		6	2	1
BM	Factor	04	09	01	20	18	02	27	31	03				
	Category	Rd	Rd	Rd	In	In	Rd	In	Re	Rd		5	3	1
LD	Factor	19	21	20	23	09	04	29	26	17	31			
	Category	In	In	In	In	Rd	Rd	Re	In	Rd	Re	3	5	2
LM	Factor	02	17	21	20	23	18	01	09	31				
	Category	Rd	Rd	In	In	In	In	Rd	Rd	Re		4	4	1
										Total		18	14	5

Factor scores in ranking phase. In phase three of the Delphi study, each panel had to rank factors according to their importance. To account for this relative importance, the rank was translated into scores ranging from 10 to 0. The highest-ranking factor in a panel got a score of 10, the second-highest 9, and so forth so that in the BD, BM, and LM panel 54 scores and in the LD panel, 55 scores could be assigned. The results are shown in Figure 32. Out of the overall distributable 217 scores, factors in the reduced influence category received 111 (51%), in the increased category 93 (43%), and in the remaining category 13 (6%). This indicates that the Guanxi impact will be reduced in the future.

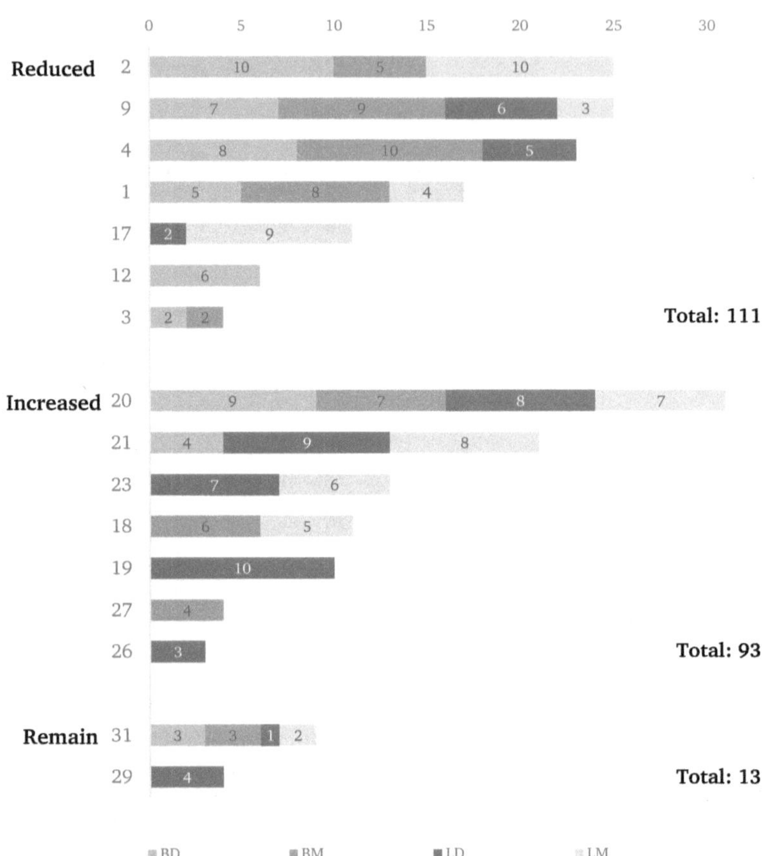

Figure 32: Scores per factor
Source: Own figure

All three indicators along the different phases of the Delphi study, the number of factors identified throughout the brainstorming session, the number of factors selected in the narrowing down phase, and the factor scores in the ranking phase indicate that the Guanxi influence will decrease in the future.

7.3.3. Group differences in future Guanxi development

In the third research question, the notion was brought forth about how factors that will impact how Guanxi's influence on the LSP selection in China will develop in the future differ among groups. Therefore, study participants were selected based on specific characteristics and divided into four panels. What follows is a summary of the results, visualized in Venn diagrams, which are commonly used for showing differences among groups in Delphi studies[1013].

LSPs compared to Buying companies

To gain insights into the perspective of LSPs compared to Buying companies, the results of the panels LD and LM (LSPs) are combined and compared to the pooled results of BD and BM (Buying companies). Factors that were selected by both panels in a group are marked with a dot ("•") on the panel name side (e.g., factor 1 was selected by the BD and the BM panel, which are both buying companies, so it is marked with "1 •". Factor 23 was selected by LD and LM, which are both LSPs, so it is marked "• 23"). This LSP group includes 13 individual factors, out of which eight are overlapping with the Buying companies. The Buying companies group includes 11 individual factors. The results are visualized in Figure 33.

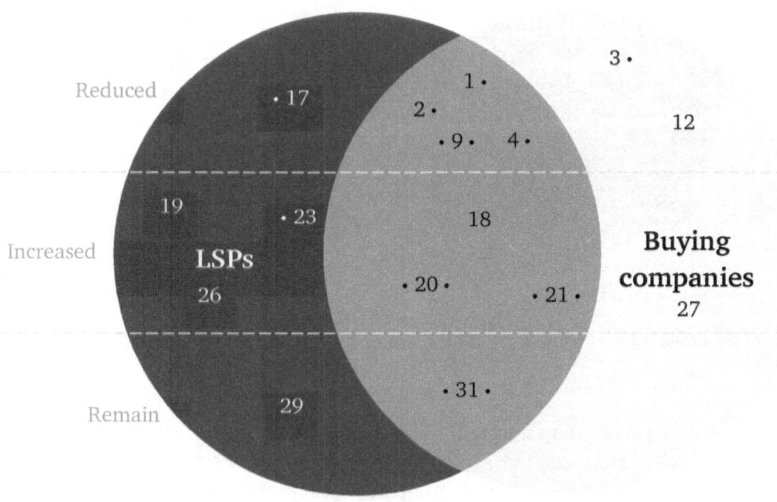

Figure 33: Factor comparison LSPs versus Buying companies
Source: Own figure

Within the LSP group, 9 factors, which means 47% of the 19 selected factors by LD and LM (including double selected ones), are part of the "increased" category. In the Buying companies group, just 5 factors, or 28% of the selected 18, are. Correspondingly, the Buying companies selected 11 factors (61%) of the "decreased" category, while the LSP group selected 7 (37%). The LSP group selected 3 (16%) of the factors from the "remaining" category, the Buying companies group 2 (11%).

[1013] Schmidt et al. (2001), p. 18.

Multinational compared to Domestic companies

To gain insights into the perspective of domestic Chinese compared to Multinational companies, the results of the panels MB and LM (Multinational) are joined and compared to the combined results of BD and LM (Domestic). Also, here, factors selected by both panels in a group are marked with a dot ("•") on the respective panel name side. The Multinational companies group contains 12 factors, out of which 10 overlap with the Domestic companies. The Domestic companies group contains 14 factors. The results are visualized in Figure 34.

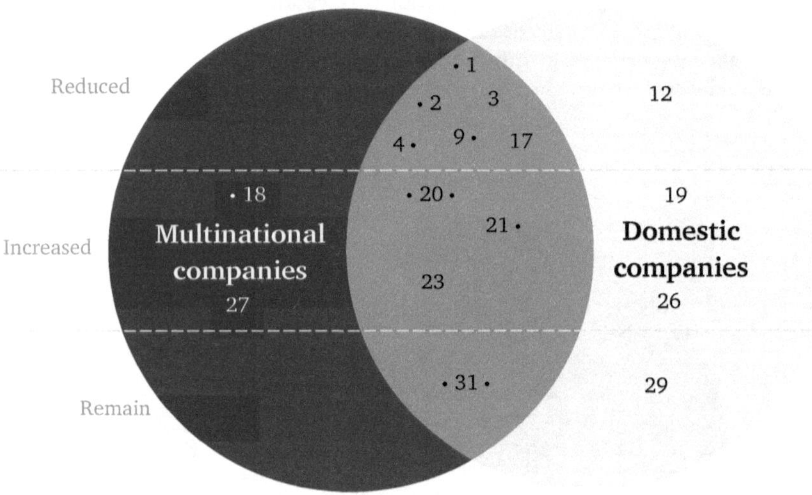

Figure 34: Factor comparison of international and Chinese corporations
Source: Own figure

The Multinational companies group selected 9 factors, or 50% of the total 18 (including double selected ones) from the "increased" category, 7 (39%) from "decreased" and 2 (11%) from "remained." Almost equally, in the Domestic companies group, 9 factors of the in total 19 are from the "increased" category (47%), 7 (37%) from "decreased," and 3 (16%) from "remained."

7.4. Discussion

The section below discusses the results of the study. First, the identified factors will be related to previous research. Then, common themes across these factors will be summarized. This is followed by a review of study limitations. Important practical implications for LSPs, Buying companies, and policymakers, as well as theoretical ones, are subsequently pointed out. The section closes by suggesting further related investigations that could be carried out.

7.4.1. Evaluation

Throughout this study, 31 factors that will impact Guanxi's future influence on the LSP selection process could be identified. They will subsequently be compared to the relevant research[1014].

The impact of Guanxi will be reduced in the future

1.   **Increased usage of artificial intelligence will lead to less influence of humans in a tender process. Guanxi is between humans.** Defining artificial intelligence (AI) is challenging due to the lack of clarity of what the term "intelligence" denotes and whether machines can possess it[1015]. A common view is that it refers to scientific studies, especially in computer science, concerned with the automation of intelligent behavior, i.e., that computers can think, do, interact and act in many fields in which traditionally people are better at[1016]. A goal of AI is to build intelligent agents, computer programs that act in an environment[1017]. In procurement, AI is hypothesized to lead to further, revolutionary automation of operative processes. Human involvement might then only be required for process management and monitoring, which would reduce the influence of human, personal relationships[1018].

2.   **Compliance will become more and more important in corporations. Guanxi might be seen as not compliant.** Although there is no universally accepted definition of corporate compliance in business research or practice, it often refers to compliance with legal and regulatory standards, including the measures needed to assure conformity with norms[1019]. These measures are "programs that corporations adopt to educate employees, improve ethical norms, and detect and prevent violations of law[1020]." Such compliance programs were increasingly set up in companies when the United States Federal Sentencing guidelines were revised and adopted "the existence of an effective compliance and ethics program" as a factor mitigating an organization's punishment[1021]. Since this 1991 revision, companies consistently developed, revised, or reinforced their ethics and compliance programs[1022]. Similarly, with the introduction of the Code of Corporate Governance in 2002, China has begun to strengthen its legal environment and to force companies into compliance[1023]. Recent surveys by the advisory firms Thomson Reuters and KPMG showed that companies will keep increasing their resource commitment to compliance activities[1024]. As laid out in section 4.6, Guanxi might be seen as not compliant and therefore be less employed in a more compliance-driven market environment.

3.   **Enforced laws and regulations against corruption will increase and Guanxi might be related to corruption.** Corruption is often defined as the misuse of public office for private gain[1025]. A case of corruption exists when "a public official (A), acting for personal gain, violates the norms of public office and harms the interests of the public (B) to benefit a third party (C)

[1014] Hasson/Keeney (2011), pp. 1695-1704.

[1015] Fetzer (1990), pp. 3-4.

[1016] Luger/Stubblefield (1997), p. 1; Rich (1985), p. 117; Rich (2004), pp. 1-55. For an overview about books discussing the nature of AI and the overlapping disciplines beyond computer science like psychology, neuroscience, philosophy, logic, and linguistics, cf. Sloman (2010).

[1017] Poole/Mackworth (2017), p. 4.

[1018] Glas/Kleemann (2016), p. 59; Pellengahr et al. (2016), p. 7; Nicoletti (2017), pp. 89-107. For a review about various themes and terms in electronic procurement cf. Glas/Kleemann (2016), pp. 55-66.

[1019] Wolf (2006), p. 1995; Schneider (2003), p. 646; Nave/Bonenberger (2008), p. 734; Rohde-Liebenau (2007), p. 275.

[1020] Baer (2009), p. 949.

[1021] Bürkle (2005), pp. 565-570.; For a review about background, core elements and impact of this review of the Federal Sentencing guidelines, cf. Murphy (2001), pp. 697-719.

[1022] Josephson (2014), pp. 13-15; Weber/Fortun (2005), p. 97.

[1023] Jiang/Kim (2015), pp. 210-214.

[1024] Thomson Reuters (2016).

[1025] Treisman (2000), p. 399.

who rewards A for access to goods or services which C would not otherwise obtain[1026]." Corruption also exists in a business setting, which becomes especially relevant in countries where the government plays a decisive role in the economy, such as in China[1027]. Especially since its economic liberalization in 1978, China is famously corruption-ridden, ranking on 80th place out of 180 considered countries in Transparency International's 2019 Corruption Perceptions Index[1028]. To tackle corruption in the country and advance the rule of law, China initiated various campaigns[1029]. The most recent one is particularly extensive and on a large-scale, sentencing already some 270,000 Chinese Communist party cadres for corrupt activities, and often seen as the boldest and most serious campaign in China[1030]. When the president launched it shortly after taking office, he warned that corruption could lead to the state's downfall[1031]. The campaign is designed according to his belief that continuous anti-corruption efforts will, over the years, deter bureaucrats from even considering corrupt behavior[1032]. Although the campaign is still ongoing, it could already be demonstrated that it significantly lowered firms' likelihood of fraud[1033]. In line with these activities, the Guanxi influence might decrease.

4. **Digitalization will make the selection processes more transparent and fair. Guanxi is mainly operating in grey areas and has unfair impacts.** Digitalization refers to the ongoing restructuring of various domains around digital communication and media infrastructures, a rethinking of current operations from new perspectives enabled by digital technology, caused by the emergence of technological platforms and information and communications technology[1034]. The base for digitalization is digitization, the conversion of analog streams of information into a digital form which can be processed by a computer[1035]. Digitalized selection processes in purchasing are enabled by e-procurement technology, a technology designed to facilitate the acquisition of goods by a commercial organization over the internet, such as e-Procurement software, B2B (business-to-business) auctions, B2B market exchanges, and purchasing consortia[1036]. In Supply relationships, transparency refers to the "two-way exchange of information and knowledge between customer and supplier[1037]." Such information sharing is beneficial for all involved members and improves the whole supply chain by, for example, reducing uncertainty and risk[1038]. It also supports to demystify "arcane, obtuse, or possibly illegal practices[1039]." For instance, in e-auctions, sellers bid for business on a level playing field rather than attempting to obtain business based on networks of personal relationships[1040]. With its informal and favor-based nature, Guanxi is sometimes considered illegal and perceived as operating in grey zones[1041]. Digitalizing procurement processes is also recommended by the Canadian Trade Commissioner Service to limit corruption risks and facilitate tracking of

---

[1026] Philp (2002), p. 42; Philp (2006), p. 45.
[1027] Murphy et al. (1991), pp. 503-530; Jain (2001), p. 97; Faure/Fang (2008), p. 203; European Commission (2017), p. 21.
[1028] Hao/Johnston (1995), pp. 80-94; Wederman (2004), pp. 920-921; Transparency International (2020), pp. 2-3.
[1029] He (2000), pp. 267-268.
[1030] Leung (2015), p. 32; Zhang (2018), p. 378.
[1031] Lin et al. (2016), p. 2; Leung (2015), p. 32.
[1032] Leung (2015), pp. 32-38.
[1033] Zhang (2018), p. 376.
[1034] Brennen/Kreiss (2016); Parviainen et al. (2017), p. 74; Corrocher/Ordanini (2002), p. 9.
[1035] Oxford English Dictionary (2017a); Brennen/Kreiss (2016).
[1036] Davila et al. (2003), p. 11.
[1037] Lamming et al. (2004), pp. 292-295.
[1038] Yu et al. (2001), p. 115.
[1039] Lamming et al. (2001), pp. 4-5.
[1040] Davila et al. (2003), pp. 13-18.
[1041] Chen, Chao C. et al. (2004), p. 200; Lee/Dawes (2005), p. 29; Li (2006), p. 105; Yang (2002), p. 461.

corruption evidence[1042]. Therefore, it can be suggested that with more transparent and fair selection processes through digitalization, the Guanxi influence will be reduced.

5. **The Chinese logistics market will become more mature and Guanxi helps to compensate market inefficiencies.** China's logistics market faces various challenges, including low efficiency and high logistics cost, congestion, lack of a nationally integrated intermodal transport network, entrenched regulation, and local protectionism[1043]. Since the decline of central planning in China in the 1980s, local governments have increasingly gained control over regulations and strongly use them to ensure local growth, employment, social stability, and tax revenues[1044]. Therefore, local governments set up barriers to protect local businesses and prevent the entry of outside competitors[1045]. In order to navigate the local regulations and ensure working logistics processes, Guanxi often is considered necessary[1046]. The central government recognized these inefficiencies and introduced structural reforms in 2015 aimed at upgrading the country's logistics network and improving logistic capacity and efficiency[1047]. Therefore, it can be assumed that with a more mature logistics market, the compensation of inefficiencies through Guanxi is not required anymore, which will translate into a reduced Guanxi influence.

6. **Women see Guanxi as less important part of business and life compared to men. Women will increasingly take over decision power.** Various scholars argue that different from contemporary Western ideals, gender definitions in the Chinese society are considered primal forces that fundamentally shape the world and cannot be separated from social roles[1048]. Well into the twentieth century, a women's place was the "inner world," consisting of family and the household, and a man's the "outer world," consisting of labor and public affairs[1049]. Therefore, China's business culture is traditionally male-dominated, and many Guanxi rituals are built on "sharing the pleasures of masculine heterosexuality" during the night together, where the provisioning of women's bodies and sexual services is used as a gift[1050]. Women are excluded from such activities and are considered to have a different approach to relationships[1051]. Women's business involvement changed in China's socialist period when they were encouraged to participate in paid work and received legal and institutional support[1052]. Although the reforms had double-sided impacts on women, the numbers of them employed in enterprises increased[1053]. Today, while still, a stark gender imbalance is visible in Chinese firms' leadership, various factors such as an increasing share of female directors on boards of Chinese firms and increased female entrepreneurs indicate a change[1054]. Women gradually take on positions that are much more in line with their capacities[1055]. Hence, it can be argued that as women see Guanxi as a less important part of business and life, the impact of Guanxi will be reduced.

[1042] Canadian Trade Commissioner Service (2018).

[1043] Zhang/Figliozzi (2010), pp. 179-190.

[1044] Jiang/Prater (2002), pp. 787-788.

[1045] Zhang/Figliozzi (2010), pp. 179-190.

[1046] Jiang/Prater (2002), pp. 785-787; Luk et al. (2008), pp. 601-609.

[1047] Jiang (2018), pp. 1-24.

[1048] Fei (1992), pp. 26-27; Stacey (1983), pp. 1-324; Anagnost (1989), pp. 313-342; Hershatter (2004), pp. 1013-1017.

[1049] Rofel (1999), pp. 1-212.

[1050] Yang (2002), p. 466.

[1051] Beetles/Crane (2005), pp. 231-233.

[1052] Shu/Zhu (2012), p. 1103.

[1053] Robinson (1985), pp. 35-38; Goodman (2002), pp. 331-333.

[1054] Liu, Yu et al. (2014), p. 174; Deng et al. (2010), pp. 3-5.

[1055] Faure/Fang (2008), pp. 202-203.

7.	**Purchasing departments will become more and more centralized and take over decision power. This makes relationship building very difficult.** An organization's structure defines roles, tasks, responsibilities, and which resources are available inside[1056]. The dominant theme regarding the organization of a company's purchasing function is to which degree to centralize or decentralize[1057]. Centralization refers to the concentration of decision-making authority or power in a single organizational unit or the position of the decision-making authority inside an organization's hierarchy[1058]. The structure of the purchasing organization impacts various needs of a company such as the integration with other areas of the organization, the leveraging of economies of scale across divisions, faster decision making by shifting decision-power to lower organizational levels, or the enabling of global sourcing by locating staff in geographically dispersed markets[1059]. Previous research showed that cost performance increases when purchasing decisions become more centralized, which might be due to benefits such as consolidation, leveraging, and the development of professional purchasing expertise[1060]. With the current Chinese market environment increasingly requiring efficiency and cost performance (also refer to the discussion of factor 8), more companies might decide to opt for a centralized purchasing structure. Purchasing centralization is typically accompanied by increased formalization and standardizing of routines. It further increases transparency regarding supplier performance, terms, and conditions[1061]. This could reduce the chances of building up close personal relationships beyond compliance regulations, as also outlined in the discussion of factor 10. The Guanxi impact would, therefore, decrease.

8.	**Profitability will become increasingly important in the future and establishing and maintaining Guanxi is expensive.** Historically, the Chinese economic system was characterized by state ownership and bureaucratic coordination through central planning, which led to high distortions and low efficiency[1062]. However, with China's opening up from 1978 onwards, China set course towards a historical era of reforms to a market economy[1063]. Throughout the past decades, privatization and restructuring of state-owned enterprises, financial system reforms, government reforms, monetary reforms, tax, and fiscal system reforms, as well as foreign-exchange reforms, were conducted[1064]. These institutional changes shaped managerial incentives, affected transaction and agency costs, and made resource allocations across and within industries selective[1065]. Local governments were incentivized to improve firm performance, which translated into low tolerance for poorly performing enterprises[1066]. Profitability pressure used to be lower in state-owned enterprises and in industries where local governments set up barriers to protect local businesses[1067]. The logistics industry, where many players are state-owned, and which is coined by inefficiencies due to local government regulations, is currently also targeted to improve its efficiency[1068]. Equally, the automotive industry in China is facing increasing cost pressure and must lower its cost

---

[1056] Robbins (1990), pp. 4-7; Jones (2013), pp. 1-25.

[1057] Johnson et al. (1998), p. 8; Glock/Hochrein (2011), pp. 155-158.

[1058] Pugh et al. (1963), pp. 289-315; Price (1972), pp. 43-57; Robbins (1990), pp. 104-113; Hickson et al. (1969), pp. 378-397.

[1059] Johnson et al. (1998), p. 3; Carter/Narasimhan (1996), pp. 20-26; Giunipero/Monczka (1997), pp. 321-335.

[1060] Ateş et al. (2018), pp. 68-78; Giunipero/Monczka (1997), p. 333.

[1061] Tate/Ellram (2012), pp. 15-24.

[1062] Kornai (1992), pp. 33-382.

[1063] Qian (2000), pp. 153-170.

[1064] Qian (2000), pp. 153-170.

[1065] Park et al. (2006), pp. 127-132.

[1066] Nee (1992), pp. 1-27; Walder (1995), pp. 263-301.

[1067] Park et al. (2006), pp. 139-146.

[1068] Jiang (2018), pp. 1-24.

structure[1069]. Generating and maintaining Guanxi is expensive and can harm profitability so that companies in the future might refrain from it[1070].

9. Purchasing processes will become more and more standardized and limit possibilities to take influence through Guanxi. The purchasing function operates with processes that link members in the supply chain and include "all activities necessary to acquire goods and services consistent with user requirements[1071]." Standardizing these purchasing processes can have a significant positive impact on a firm's profitability[1072]. The target of process standardization is, among others, to reduce costs by decreasing process errors, facilitating communication, or just incorporating expert knowledge[1073]. China's logistics and automotive industry are currently undergoing a transformation towards increased profitability and efficiency, requiring the adoption of standardized processes (cf. the discussion about factor 8). Organizational purchasing behavior refers to the behavior of individuals that is constrained by policies, such as standardized processes[1074]. Such defined processes could reduce the risk for unintended personal influences, for example, through the application of certain Guanxi processes[1075].

10. Efficiency will become increasingly important in the future and establishing and maintaining Guanxi is time intensive. *It was clarified with the experts in the brainstorming session that "efficiency" refers to the efficient usage of the working time of the boundary spanners typically involved in Guanxi building, such as sales departments.* In economic theory, efficiency refers to outputs that cannot be improved without making something else worse (Pareto efficiency)[1076]. The efficiency of production units refers to comparing observed and optimal output and input value, which is closely linked to productivity, the ratio of output to its input[1077]. Adopted to the production factor labor, it refers to the optimal ratio of a department's output to workforce input, typically in the form of effective work time[1078]. Therefore, allocating the workforce's time on activities that are not adding to the expected department output would reduce a department's efficiency. Developing Guanxi is time-intensive, and only few Guanxi relationships can be maintained[1079]. Therefore, depending on the target of various departments and the positive as well as negative effects of Guanxi, e.g., increasing market share but lower profitability, employing Guanxi can lead to lower efficiency[1080]. However, as discussed in factor 8, the logistics and automotive industry are currently targeted to improve their efficiency, so that the Guanxi usage might be reduced[1081].

11. Younger generations (especially post-90) see Guanxi as less important part of business and life compared to previous generations. These younger generations will increasingly take over decision power. Various changes in people's values could be observed across different generations [1082]. Especially changes in the economic environment were

---

[1069] Thun (2018), pp. 7-18.

[1070] Park/Luo (2001), p. 462; Warren et al. (2004), pp. 355-364; Li et al. (2009), pp. 347-348.

[1071] Novack/Simco (1991), p. 145.

[1072] Sánchez-Rodríguez et al. (2006), p. 62; Bennett (1982), pp. 41-42.

[1073] Toni/Panizzolo (1993), pp. 1379-1381; Wüllenweber et al. (2009), pp. 527-548; Manrodt/Vitasek (2004), pp. 1-23.

[1074] Webster Jr/Wind (1972), p. 18; Schäfermeyer et al. (2010), p. 2.

[1075] Anand et al. (2004), pp. 49-50; Lee/Dawes (2005), p. 29; Wells (2014), pp. 42-48.

[1076] Acemoglu et al. (2016), p. 180.

[1077] Lovell (1993), pp. 3-4.

[1078] Pindyck/Rubinfeld (2018), p. 212; Deprins et al. (2010), pp. 298-299.

[1079] Fan (2002b), p. 547; Yi/Ellis (2000), pp. 28-29.

[1080] Park/Luo (2001), p. 462; Li et al. (2009), pp. 347-348.

[1081] Thun (2018a), pp. 7-18; Jiang (2018), pp. 1-24.

[1082] Inglehart (2008), p. 130; Loughlin/Barling (2001), p. 543; Thornton/Young-DeMarco (2001), p. 1009.

accompanied by changes in the values of younger generations[1083]. Naturally, with China's transition towards a market economy, such changes could also be expected[1084]. Indeed, studies showed value shifts in China over the past decades, such as increased individualistic values[1085]. Some scholars consider the phenomenon of Guanxi and the corresponding values as dynamic[1086]. Also, studies showed that younger generations resort less to Guanxi and relate it instead to incompetence in a work setting[1087]. Over time, this generation will enter the workforce and take over organizational control, so that it can be established that the Guanxi influence will decrease.

12.   Logistics will become a more important function in a company in the future. In functions with high focus Guanxi impact is actively reduced. Logistics is currently undergoing a transformation with increasing costs for shipping of goods and services, technological changes, and rising customer expectations towards faster, more flexible delivery at lower costs[1088]. Issues and trends become strategic when top management believes that it has relevance for organizational performance[1089]. With increasing relevance for organizational performance, top management raises its attention towards a company's function, as, for example, observed with purchasing[1090]. This also includes monitoring of possible managerial malfeasance and setting up measures guarding against it[1091]. It could be demonstrated that the perceived behavioral control, that is, the perception of the ease or difficulty of performing the behavior of interest, influences the intention to conduct unethical behavior[1092]. Some Guanxi practices can be considered such, so it can be assumed that the Guanxi impact will be reduced with growing importance of the logistics function.

13.   Cooperation and communication become less dependent on relationships between individuals. Guanxi is related to the relationship between individuals. *It was clarified with the experts in the brainstorming session that "cooperation and communication" refer to cooperation and communication of supply chain partners.* Robotic Process Automation (RPA) refers to tools that operate on the user interface of other computer systems in ways humans would[1093]. They aim at reducing the burden of repetitive, simple tasks on employees[1094]. Chatbots are computer programs that can hold conversations in a natural language speech with humans[1095]. They can serve several purposes, including customer service and information sharing[1096]. These technologies are highly efficient and cost-effective, so that they have high potential to reduce the involvement of individuals across a supply chain[1097]. Automation of operational tasks will change the nature of communication among boundary spanners who might then only focus on value-adding activities[1098]. Such developments could reduce the impact of Guanxi in the future.

[1083] Flanagan (1982), pp. 434-436.
[1084] Park et al. (2006), pp. 127-132.
[1085] Zeng/Greenfield (2015), p. 53.
[1086] Lovett et al. (1999), p. 237; Luo et al. (2012), pp. 148-152.
[1087] Faure/Fang (2008), p. 197; Hanser (2002), pp. 160-161; Chen et al. (2013), pp. 193-194; Wilson/Brennan (2010), pp. 662-663.
[1088] Tipping/Kauschke (2016), p. 3; Stank et al. (2015), pp. 28-31; Council of Supply Chain Management Professionals (2018), pp. 1-72.
[1089] Dutton/Ashford (1993), p. 397.
[1090] McIvor et al. (1997), pp. 176-177.
[1091] Adams et al. (2010), pp. 65-82.
[1092] Park/Blenkinsopp (2009), pp. 546-548; Chang (1998), pp. 1829-1832; Ajzen (1991), p. 181.
[1093] Van der Aalst et al. (2018), p. 269.
[1094] Aguirre/Rodriguez (2017), p. 65.
[1095] Sameera (2015), p. 72.
[1096] Brandtzaeg/Følstad (2017), 377-381.
[1097] Van der Aalst et al. (2018), p. 269; Aguirre/Rodriguez (2017), p. 65.
[1098] McIvor et al. (2003), p. 150; Nicoletti (2017), pp. 89-107.   154

14. Internationalization of the logistics market will lead to an adoption of international business practices. Business practices are methods, procedures, processes, or rules employed or followed by companies to pursue their objectives[1099]. There is debate among researchers if the internationalization of businesses will lead to homogeneity, polarization, or hybridization of business practices[1100]. This becomes especially controversial when attempting to operationalize implications of internationalization by, e.g., trying to develop internationally valid business codes of ethics[1101]. Regardless of this controversy, some countries already passed laws that impact companies not only operating in their own country but internationally[1102]. A further internationalized Chinese logistics market will, therefore, also require compliance with such laws and regulations. Some of the practices close to Guanxi might be perceived as violating international laws (cf. section 4.6). Also, Guanxi might be considered as a phenomenon that is originating from the lack of market system trust during China's transition[1103]. With gradually increasing internationalization, the Chinese logistics market might mature, hence rendering Guanxi less important[1104].

15. Managers see Guanxi as less important part of business and life compared to before. A school of thought believes that Guanxi is the Chinese form of the personal relations required to accomplish tasks in transforming economies[1105]. Therefore, as the Chinese economy and especially the logistics market are now maturing, the practice of Guanxi might become obsolete[1106]. In local business practices, a tendency can be observed where Guanxi alone is not sufficient for long-term success and where product/service performance is becoming more imperative[1107]. For instance, studies among managers in Shanghai showed a tendency that law becomes more significant than the art of Guanxi practice[1108]. Some scholars further recognize Guanxi as a typical social capital form occurring in collectivist cultures[1109]. With changing values towards individualism in China, this collectivist "we consciousness" might likewise decrease[1110].

16. Associate's turnover will increase in the future. It will be therefore more difficult to build up Guanxi relationships. Employee turnover rates in China have been rising over the past years. From similarly low turnover rates to the US and Japan in the 1980s, they increased to 20.8% in 2016[1111]. Developing Guanxi is time-intensive, and only a few Guanxi relationships can be built up and maintained[1112]. Therefore, with further rising turnover rates, it might become increasingly difficult to build up Guanxi with the right stakeholders inside a company.

17. The reputation of a corporation becomes increasingly important in the future. Using practices related to corruption harms a corporation's reputation when discovered. Using Guanxi is sometimes connected with corruption. For LSPs, reputation refers to the customer's opinion about how good the LSPs are at satisfying customer's needs[1113]. An LSP's image in the industry is one of the key criteria for selecting Logistics Service Providers[1114]. While

---

[1099] WebFinance (2019).
[1100] Asgary/Walle (2002), pp. 69-70.
[1101] Asgary/Mitschow (2002), pp. 240-245.
[1102] Asgary/Mitschow (2002), p. 240.
[1103] Berger/Herstein (2012), pp. 29-30; Shi et al. (2011), p. 496; Hwang (1987), p. 956.
[1104] Tan et al. (2009), p. 544.
[1105] Walder (1988), pp. 1-254.
[1106] Tan et al. (2009), p. 544; Jiang (2018), pp. 1-24.
[1107] Wiegel/Bamford (2015), p. 317.
[1108] Guthrie (1998), pp. 281-282.
[1109] Gu et al. (2008), p. 15.
[1110] Parsons/Shils (2017), p. 82; Hofstede (2001), p. 227; Zeng/Greenfield (2015), p. 53.
[1111] Ma/Trigo (2008), p. 35; Aon plc (2016).
[1112] Fan (2002b), p. 547.
[1113] Aguezzoul (2014), pp. 69-78.
[1114] Alkhatib et al. (2015), p. 133; Hwang et al. (2016), p.

remaining among the most important criteria for decades, the relative importance of reputation has been increasing over time[1115]. Reputation damage can, therefore, have adverse effects on an LSP's future business prospects in an increasingly competitive market[1116]. When discovered, corrupt behavior can risk a company's reputation and raise doubts about its honesty in other transactions[1117]. Some of the Guanxi practices are seen as related to, or even perceived as an integral part of corruption in China[1118]. Therefore, with the increasing importance of a company's reputation, resorting to Guanxi might be reduced in LSP selection processes.

The impact of Guanxi will be increased in the future

18.   The business environment is becoming more volatile, uncertain, complex and ambiguous. Guanxi helps to reduce risks in such a business environment. Volatility refers to frequent and sometimes unpredictable change. Uncertainty is a lack of knowledge as to whether an event will create significant change. Complexity means that many interconnected parts form an elaborate network of information and procedures that is often multiform and convoluted. Ambiguity is a situation where cause and effect are not understood, and there is no precedent for making predictions as to what to expect[1119]. These attributes, often abbreviated VUCA, lead to increasing supply chain risks[1120]. As it is trust and favor exchange based, Guanxi can be used as a strategy for mitigating supply chain risks[1121]. Supply chain partners need to support each other when in need, information is openly shared, and actions are more likely jointly coordinated[1122]. In an increasingly VUCA business environment, it can be argued that the impact of Guanxi will increase.

19.   Contractees will use more strategic purchasing in the future. Characteristic for strategic purchasing is a closer relationship with suppliers and Guanxi is about building relationships. Purchasing is increasingly considered a strategic function and recognized for its positive contribution to a firm's performance[1123]. The function can support to achieve high quality levels, fast delivery, cost savings, and strategically acquire needed resources. With further raising customer demands, it is foreseeable that more firms will adopt strategic purchasing in the future[1124]. A key element of strategic purchasing are long-term supplier relationships, and Guanxi can be a measure to build these up[1125]. It can thus be assumed that the Guanxi impact will increase.

20.   Services will become more customized in the future and mutual understanding through information exchange is the base for service customization. Guanxi supports information sharing between business partners. To satisfy further increasing heterogeneous customer needs, individualization is needed in the form of further product customization[1126]. Logistics is an important part of this customization, especially when manufacturing is postponed in the distribution channel close to the customer. This already required more service

---

[1115] Weber et al. (1991), pp. 2-18; Ho et al. (2010), pp. 16-24. A recent study could not further confirm this upwards trend Alkhatib et al. (2015), p. 111.

[1116] Huo et al. (2015), p. 163.

[1117] Argandoña (2005), p. 258; Luo (2005), pp. 139-141; Zekos (2004), pp. 632-634.

[1118] Zhan (2012), pp. 99-105; Li (2011), p. 20.

[1119] Bennett/Lemoine (2014), p. 313.

[1120] Doheny et al. (2012).

[1121] Abramson/Ai (1999), pp. 21-31; Cheng et al. (2012), p. 11.

[1122] Lee/Humphreys (2007), pp. 462-463; Sheu et al. (2006), pp. 25-46; Barnes et al. (2011), p. 512.

[1123] Paulraj et al. (2006), p. 109.

[1124] Carr/Pearson (2002), pp. 1032-1035.

[1125] Paulraj et al. (2006), p. 109; Cai et al. (2017), p. 19; Geng et al. (2017), pp. 10-11; Luo et al. (2012), p. 140.

[1126] Brettel et al. (2014), pp. 37-38.

customization from LSPs[1127]. In the LSP literature, one school of thought argues that this service customization is the next evolutionary step for LSPs[1128]. Providing customized services requires client participation, primarily through the provision of information[1129]. Guanxi supports information sharing among business partners, especially when an atmosphere of trust is needed, such as with service customization[1130]. This supports the idea that the Guanxi impact is going to increase.

21. **Trust is the basic foundation of business and will become more important. Guanxi supports building up trust.** Trust reduces the perception of risk associated with opportunistic behaviors, increases confidence that short-term inequities will be resolved over a long period, and reduces transaction costs in an exchange relationship[1131]. Therefore, it is precondition to ensure long-term oriented business relationships[1132]. Especially in relation-based societies such as China, it is a critical foundation to conduct business[1133]. As the Chinese logistics market is maturing, outsourcing logistics to LSPs is expected to become more common[1134]. Accordingly, companies will need to set up new supplier relationships. A defining element of Guanxi is trust, and Guanxi is known to support its build-up, so that its importance might increase in the future[1135].

22. **Profitability will become increasingly important in the future and Guanxi can lead to competitive advantages.** As described in the discussion about factor 8, the logistics and automotive industry are currently targeted to improve their efficiency[1136]. Competitive advantage refers to superior customer value a company can provide by either offering lower prices than competitors for equal benefits or providing unique benefits that justify higher prices[1137]. Guanxi can be considered a resource for competitive advantage as it is an intangible asset that is semi-permanently tied to the firm and leads to higher performance[1138]. For instance, Guanxi benefits include gaining valuable and relevant information, enhancing business, and decreasing transaction costs[1139]. With increasing importance of profitability and a correspondingly further push for the sharpening of competitive advantages, Guanxi can become more important.

23. **Service quality will become increasingly important in the future. An important element of Guanxi is the exchange of favors beyond contractual agreements which will improve the service quality.** Logistics is increasingly regarded as a powerful source of competitive differentiation to address the challenges of a market environment coined by faster-changing information technology and increasing diversification of consumer requirements[1140]. Service quality in logistics refers to timeliness, condition and accuracy of the order, quality of information, and contact personnel's availability and quality[1141]. It is positively impacted by logistics flexibility, i.e., adjusting capacities and schedules if customer demand is fluctuating,

---

[1127] Van Hoek (2000), p. 37.
[1128] Skjoett-Larsen et al. (2006), pp. 194-195; Murphy/Poist (1998), p. 26.
[1129] Maglio/Spohrer (2008), p. 18.
[1130] Wiegel/Bamford (2015), p. 315; Shin et al. (2007), pp. 171-172.
[1131] Ganesan (1994), p. 3.
[1132] Chumpitaz Caceres/Paparoidamis (2007), p. 846.
[1133] Yeung/Tung (1996), p. 63.
[1134] Chu/Wang (2012), pp. 78-79.
[1135] Chen/Chen (2004), pp. 313-314; Kriz/Keating (2010), pp. 308-309.
[1136] Thun (2018), pp. 7-18; Jiang (2018), pp. 1-24.
[1137] Porter (1998), p. 3.
[1138] Wernerfelt (1984), p. 172; Yeung/Tung (1996), p. 63; Lee et al. (2018), pp. 357-358
[1139] Shin et al. (2007), pp. 171-172; Luo et al. (2012), pp. 158-161; Standifird/Marshall (2000), p. 30; Wang (2007), p. 85.
[1140] Fredericks (2005), pp. 562-563; Richey et al. (2007), pp. 197-198 Mentzer et al. (2001), p. 82 Mentzer et al. (2004), 15-19; Yu et al. (2017), p. 211.
[1141] Parasuraman et al. (1988), pp. 15-36.Gil Saura et al. (2008), p. 653

and relationship flexibility, the expectation of being able to make adjustments in the ongoing relationship when an unexpected situation arises[1142]. A Guanxi relationship involves the exchange of favors and support when required by the other party. It can include factors that are part of logistics and relationship flexibility[1143]. The increasing importance of service quality might, therefore, be reflected in an increasing future Guanxi influence.

24. Selection processes will become increasingly more efficient in the future. Guanxi leads to a more efficient information exchange. Logistics Service Provider selection processes are crucial, considering that with the right provider, values of logistics services, suppliers, and customers can be improved[1144]. Various tangible and intangible criteria must be considered during the selection process, which leads to high complexity[1145]. A variety of methods, such as multiattribute decision-making techniques, statistical approaches, or mathematical programming, can be used to address this challenge[1146]. It is predicted that especially those methods that increase the purchasing decision's efficiency will be increasingly used[1147]. Information exchange can reduce information asymmetry, improve decision transparency, further reduce partners' behavioral uncertainty, and aid in the collaboration of parties involved[1148]. The prevalence of Guanxi supports information sharing among potential business partners[1149]. Therefore, it can indicate that the Guanxi impact will increase in the future.

25. Logistics outsourcing will become more important in the future as it enables companies to concentrate on core competencies. Guanxi is about relationships and supports outsourcing. Strategic outsourcing enables companies to focus their limited resources on core competencies[1150]. Outsourcing improves company performance as it can lead to increased flexibility and economies of scale benefits[1151]. Globally, logistics is among the most outsourced functions[1152]. While in developed markets, the logistics outsourcing rate is above 40%, it is only 3% in China[1153]. With the currently ongoing maturing of China's logistics market, this rate is predicted to increase[1154]. Guanxi can support outsourcing as it emphasizes relationships and helps to reduce transaction costs[1155]. Consequently, it can be expected that the Guanxi impact will increase.

26. Contractor's reputation will become a more important decision criterion in the future. Companies with high Guanxi have a higher reputation. Reputation refers to the opinion of customers about how good an LSP satisfies their needs. During a selection process, this is especially relevant in the initial screening phase for potential partners[1156]. Reputation is already among the most important criteria when selecting Logistics Service Providers[1157]. Over time the importance of individual selection criteria changes. For the past years, especially the formerly highest criteria, "costs" and "price" are in relative decline[1158]. Therefore, an LSP's

---

[1142] Yu et al. (2017), pp. 217-223.
[1143] Lee/Humphreys (2007), pp. 462-463; Cheng et al. (2012), p. 7.
[1144] Alkhatib et al. (2015), p. 112.
[1145] Aguezzoul (2014), p. 71.
[1146] Chai et al. (2013), pp. 3874-3881.
[1147] Carter, Phillip L. et al. (2000), pp. 14-26; Ogden et al. (2005), pp. 29-48; Boer et al. (2001), pp. 75-76.
[1148] Chu/Wang (2012), p. 84.
[1149] Sheu et al. (2006), pp. 25-46; Barnes et al. (2011), p. 512; Shin et al. (2007), pp. 171-172.
[1150] Quinn/Hilmer (1994), p. 43; Kremic et al. (2006), p. 469; Boyson et al. (1999), pp. 73-100.
[1151] Bustinza et al. (2010), pp. 286-287; Cachon/Harker (2002), p. 1314; Grossman/Helpman (2002), pp. 85-120.
[1152] Payaro/Papa (2017), p. 46.
[1153] Zhu et al. (2017), pp. 29-30.
[1154] Wang (2018), pp. 71-93.
[1155] Lee/Humphreys (2007), pp. 454-462; Davies et al. (1995), pp. 208-213.
[1156] Maltz (1995), pp. 73-80.
[1157] Aguezzoul (2014), pp. 69-78.
[1158] Alkhatib et al. (2015), p. 111.

reputation becomes a relatively more important selection criterion. Guanxi, for example with government officials, supports building up a company's reputation and image in the minds of key stakeholders[1159]. Owing to this, it can be assumed that the Guanxi impact will increase.

27. **Market shares will become increasingly important in the future. Guanxi leads to willingness to invest into the growth of partners.** Although China's logistics market is still very fragmented, signs of consolidation are visible[1160]. With an increase in logistics outsourcing activities in China, LSPs need to keep market shares to remain competitive[1161]. This becomes especially crucial with increasing logistics cost pressure due to China's economic slowdown and logistics market reforms[1162]. Guanxi supports the closer collaboration of partners and relationship-specific investments[1163]. This can indirectly increase market shares. Besides, it was shown that Guanxi can support enhancing business and has a positive impact on firms' market shares[1164]. As a result, it can be inferred that the Guanxi impact will increase.

28. **The profitability pressure will be higher in the future. This might lead to a market consolidation including the bankruptcy of contractors. To survive as contractor a strong relationship with the contractee is important and Guanxi is about relationships.** In the past years, especially rising wages and increasing warehouse rental costs led to a decline in LSPs' profit margins. While the average profitability dropped from 6.7% in 2014 to 6.1% in 2015, some companies were already facing losses[1165]. This trend is expected to continue throughout the next years, and market consolidations are expected[1166]. LSPs and their clients are ideally connected through a relationship that goes beyond an ordinary buyer-seller relationship in the form of a strategic partnership[1167]. Guanxi is about building and maintaining such close relationships that can lead to positive business performance[1168]. Thus, it can be presumed that the Guanxi impact will increase.

The impact of Guanxi will remain the same in the future

29. **Guanxi will always be an essential part of the Chinese culture.** According to some scholars, Guanxi is a deeply rooted part of Chinese culture[1169]. While Guanxi adapts to new social institutions and structures, it remains an essential part of Chinese culture[1170]. It can be therefore assumed that the impact of Guanxi will remain the same in the future.

30. **No future factor will affect the impact of Guanxi on the selection process.** Gold argues that Guanxi, as a defining element of Chinese culture, has existed for thousands of years and is handed down relatively unchanged through time and space[1171]. Fan also postulates that certain types of Guanxi will remain largely unchanged over a long period of time[1172]. Therefore, factors, such as the previously mentioned ones concerning the logistics market and the actors within, might not alter Guanxi's impact on the selection process. It will thus remain the same in the future.

[1159] Davies et al. (1995), p. 211; Lee et al. (2018), pp. 357-358.
[1160] Wee Kwan Tan et al. (2014), pp. 1331-1332.
[1161] Wang (2018), pp. 71-93.
[1162] Jiang (2018), pp. 1-24.
[1163] Grover et al. (2002), pp. 233-236.
[1164] Peng/Luo (2000), p. 494.
[1165] Liu (2018), p. 178.
[1166] Liu (2018), p. 178; Wee Kwan Tan et al. (2014), pp. 1331-1332.
[1167] Wang (2018), p. 72.
[1168] Sheu et al. (2006), pp. 25-46; Barnes et al. (2011), p. 512; Luo et al. (2012), pp. 158-161.
[1169] Chen et al. (2013), p. 193.
[1170] Yang (2002), p. 459.
[1171] Gold et al. (2002), p. 3; Chen et al. (2013), p. 193.
[1172] Fan (2002b), p. 553.

31.   Personal feelings will always remain in business and Guanxi is related to personal feelings. Numerous studies have shown that personal feelings have an impact on businesses around the world. Among others, personal feelings can impact motivation, risk management, investment, and purchasing decisions[1173]. Also, personal bias was observed in supply chain management[1174]. Following previous investigations about the human brain's functionality, it can be assumed that at least to a certain degree, personal feelings will always remain in business[1175]. An essential element of Guanxi is an emotional attachment or good report between actors[1176]. Und this consideration, it can be concluded that the Guanxi impact will remain the same.

Implications

This study's findings have several important implications for the future *practice* of LSPs, Buying companies, and policymakers. For *LSPs*, a variety of opportunities and threats arises. With increasing compliance regulation and enforcement, the impact of the "dark side of Guanxi" will be reduced. This provides an opportunity for "fair playing" LSPs to gain new customers that were previously inaccessible. Empowered, skilled supply chain departments could be free to select the most competitive LSP, which results in increased market opportunities for them. However, this also becomes a risk for a service provider that used to operate on a Guanxi based business model as compliance regulations and transparency will transform Guanxi reliance from an asset into a liability. The new competitive environment will require LSPs to offer more customized services and higher service quality to secure increasingly important market shares. To this, Guanxi can help to build up the required trust and facilitate information sharing. Hence, LSPs should consider to strategically leverage personal relationships with key stakeholders at current and potential customers. This is also in line with the rise of strategic purchasing and reliance on trust. Together with the development and digital transformation of the supply chain function, LSPs will have to keep ahead of their customers, which will require investments into ecosystems, skills, and organizational changes. They must also find new ways how to build up and maintain Guanxi with weaker personal links on the operational level. Finally, multinational LSPs could rethink their Guanxi approach with government officials. It might allow them to gain access to privileged information so that they can mitigate the effects of an increasingly VUCA environment in China. It would also positively reflect their reputation, thus helping them gain a competitive edge in less price-driven tenders. Furthermore, there are important implications for *Buying companies*. China's push for organizational compliance means that they must implement functional compliance systems and set up compliance officers that understand how to enforce them. The results of the previous studies (cf. chapters 5 and 6) could help them to understand where Guanxi can play a role in the selection process for LSPs so that they can take appropriate countermeasures if needed. At the end of 2019, a deputy director at the Commercial Legal Service Center of China Council for the Promotion of International Trade already indicated that China would sanction non-compliant companies[1204]. Stronger regulation and more transparency in the selection process are a chance for buying companies to improve their logistics performance lastingly. When LSPs with a purely relationship-based business model vanish, the quality of the logistics market in China will progress. Also, relationship costs, such as bribes, will not be channeled to rent seeking buyers anymore, allowing LSPs to improve their price offer towards a contractee. At the same time, higher customer requirements will force buying firms to build up Guanxi with selected LSPs strategically. They can become close, trusted partners who are willing to commit to the buyer-supplier relationship and take long term investments, resulting in higher service quality offerings tailor-made to the buying company's requirements. In this market environment, companies that previously did not consider logistics outsourcing might feel confident enough to contract with LSPs. Strategically building up Guanxi with companies and government officials is also recommended to counter the impacts of a VUCA market environment. Supply chain risks can be mitigated, and crucial information shared timely. There are also implications for *policymakers*. The Chinese government should keep up their efforts to provide a market environment according to international standards, where companies can compete equally. Although most study participants perceived positive changes, the LD panel choosing only one compliance-related factor on a very low rank could indicate that the efforts are not yet comprehensive enough. Policymakers should also consider what the changes mentioned above would mean for their country's labor force. Specifically, how can they prevent skill obsolescence and facilitate the transition towards a more sophisticated and digitalized logistics market. The study findings also have significant implications for the *theoretical* understanding of the intricacies of Guanxi. Mainly due to a strengthening of compliance and the development and digital transformation of the supply chain function, the overall importance of Guanxi is expected to decrease. This is in line with the argumentation of a school of thought that predicts that with a change of structural and institutional conditions, notably the emergence of a more rational-legal system, the practice of Guanxi will become obsolete[1205]. Enhanced domestic laws, and international pressure towards more compliance, require buying companies and service providers to implement and follow stricter regulations[1206]. In contrast, while these factors indicate a reduction of the future Guanxi impact, others indicate the opposite. Market shifts in the form of increasing volatility, service customization, and service quality require closer cooperation models. Those depend on mutual support, open information sharing, trust, and relationship-specific investments. Taken together, the results suggest that Guanxi will not disappear, but instead evolve. Its dark, noncompliant side will be left behind, whereas its positive elements prevail. Following Tang's understanding, Guanxi will, therefore, remain as an ever-developing set of practices that keeps transforming and adapting over time[1207].

Future research

In terms of directions for future research, several paths are worth exploring. Considering that Guanxi is a phenomenon of the Chinese culture, the study was conducted there. Further studies could clarify how personal relationships in the selection of LSPs will develop in cultures that

---

[1204] Xinhua (2019).

[1205] Guthrie (1998), p. 273; Xin/Pearce (1996), p. 1654

[1206] Jiang/Kim (2015), pp. 210-214; Gugler/Shi (2009), pp. 3-24; It is worth noting that in comparison to the other panels, the LD panel selected only one compliance related factor, and only on the second lowest rank. As many domestic logistics companies are government related, their predictions might support the reasoning of Yao, according to which certain privileges are granted and still protected by the political system. Cf. Yao (2002), p. 293.

[1207] Yang (2002), pp. 459-460.

are either more distant, like the Western one, or closer and with similar concepts such as the Japanese or the Korean one[1208]. Results could lead to a better understanding of the influence of market conditions, political systems, and cultural characteristics on BSRs. This study's panels were set up according to firm type and internationalization degree, which enabled the identification of nuanced differences among groups. Acknowledging the findings from the studies laid out in chapters 5 and 6, also other panels could be set up for future research. For instance, panels based on different firm sizes or degree of selection process structuredness might lead to new insights and a deeper understanding of how Guanxi's influence on the LSP selection in China will develop in the future. As this study was conducted with participants working in the automotive industry or at their LSPs, findings might be biased towards industry characteristics. Applying a similar research design and conducting the study with experts from other industries, such as the fast-moving consumer goods (FMCG) or the e-commerce sector, could help to distill industry independent results. Moreover, various factors that likely impact the future Guanxi influence were identified. While they were reviewed under consideration of the current literature and explained by experts, further work is needed to understand their implications fully. For example, how will the nature of BSRs change when more software tools will handle routine work for operational staff. Finally, as China is a dynamically changing country and is affected by political developments, the study could be repeated in a few years. Advances in this field could be identified, and the impact of programs such as the Belt and Road Initiative or the New Foreign Investment Law measured[1209].

[1208] Hitt et al. (2002), p. 358; Yang/Wang (2011), p. 492; Usunier/Lee (2005), p. 464; Yeung/Tung (1996), p. 63.
[1209] The World Bank Group (2018a); Zhao (2019), p. 351.

163

8. Conclusion

As illustrated in Figure 35, this chapter marks the end of this work.

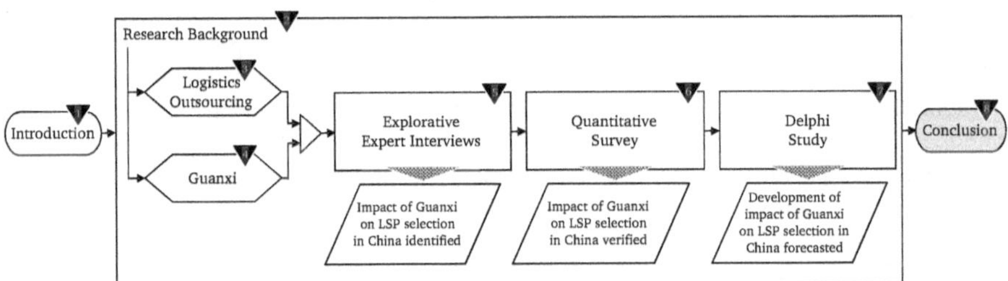

Figure 35: Chapter "Conclusion" within overall research process
Source: Own figure

Throughout this conclusion, the research results of this study will be summarized. In addition, implications will be given under consideration of the study context and the motivation that was laid out in the beginning.

[1210] Chu et al. (2018), p. 291; Rahman et al. (2019), pp. 607-616; Pagell et al. (2005), pp. 371-394; Chen et al. (2010), pp. 279-280; Chu et al. (2019), pp. 620-630.

[1211] Alkhatib et al. (2015), p. 105; Aguezzoul (2014), pp. 69-78.

[1212] Armstrong/Shimizu (2007), pp. 966-981.

[1213] Meuser/Nagel (1991), pp. 455-466.

8.2. Implications

The findings from these studies contribute in several ways to our understanding of Guanxi practices in supply chain management. For the first time, it could be demonstrated how Guanxi

---

[1214] Schmidt et al. (2001), pp. 7-13; Schmidt (1997), pp. 768-774.

impacts the selection of LSPs in China. This appears to be also one of the first comprehensive investigations into the workings of Guanxi in SCM on a micro-level. For instance, employees of a buying center share information about upcoming projects before the official start, evaluate a bidder's quality more positively, or consider bidders as possible bidder. These findings also extend the literature about BSRs during the procurement of intangible, heterogenous services such as third-party logistics. Personal relationships can positively impact a buyer's quality evaluation and lead to an information advantage for service providers. At the same time, limitations of personal relationships were documented, so that it could be shown that buyers seldom share commercially highly sensitive information. The study also makes a noteworthy contribution to the existing knowledge about contingency factors that mitigate the Guanxi impact. Identified factors include a company's internationalization degree and size, as well as a decision makers' home country. The data also suggests that neither a decision makers' age nor sex has an impact. An important theoretical implication is also that Guanxi will not disappear, but instead evolve. This adds to the school of thought that argues that Guanxi is subject to structural and institutional conditions. Here, the strengthening of compliance and the development and digital transformation of the supply chain function are predicted to reduce, while the general market shifts are predicted to increase the impact of Guanxi.

The study findings also suggest several courses of action for practitioners. Above all, *buying companies* should acknowledge that Guanxi relationships, respectively personal relationships in an international context, have certain effects on the selection of LSPs. This allows for deliberate decision making regarding their management. Not taking appropriate measures might lead to the selection of less qualified LSPs at higher prices or reputation loss. It is especially relevant for buying companies in higher-risk groups, such as SMEs, companies that mainly operate domestically, and companies of Chinese origin. Companies should consider avoiding and mitigating effects by implementing relevant guidelines and processes, in addition to improving the awareness of employees. Associates must understand better at what stages risks for unfair tender processes exist, such as bid-rigging, which occurs already before the start of a project. This becomes particularly pressing as China is increasing its push for the implementation of functional compliance systems while sanctioning non-compliant companies. Buying companies can also start to use Guanxi purposefully. Increased requirements from their customers will force them to cooperate closer with selected LSPs. They must be able to rely on partners committed to a long-term BSR and invest into assets, people, and technology that enable customized, high-quality services. Leveraging Guanxi with business partners and government officials can also contribute to alleviating impacts of a VUCA market environment. Inherent supply chain risks could be mitigated through instantaneous sharing of vital information. There are also important practical implications for *LSPs* who should use Guanxi strategically to benefit from its business-enhancing effects. Personal relationships are especially in China an essential element when contractees decide for an LSP. LSPs must therefore investigate who in their organization is directly or indirectly part of the Guanxi network of key persons from their potential clients. If they do not have an associate who is, they could deliberately hire salespeople with the right relationships. LSPs could also enhance their key account managers' relationship building skills by learning from proven Guanxi practices. As the findings imply, crucial advantages of such relationships are an enhanced sharing of commercially sensitive information, being considered as a potential bidder, as well as better quality evaluations. LSPs should also seize the chance of the changing market environment. The increasing quality of logistics services in a more competitive Chinese market might give industrial clients the required confidence to outsource more logistics tasks. Besides, as illegitimate Guanxi practices are expected to be reduced, a new level playing field will arise where business from clients that was before unattainable might become accessible. Likewise, LSPs that have been relying on illicit practices to acquire business will need to redefine their competitive positioning urgently.

8.3. Future research

Moving forward, there are additional opportunities for future research. The studies were conducted with firms in the automotive industry, which is recognized for its complex, sophisticated, and continuously improving supply chain[1215]. Therefore, interindustry effects could be avoided, but it creates a risk for a potential bias towards industry characteristics. To develop a full picture of the Guanxi impact on the LSP selection in China, additional studies in other industries are needed. For example, differences might appear in the e-commerce or FMCG industry, where logistics' main purpose is final customer distribution instead of production supply. Future investigations should also be carried out in different countries to develop a deeper understanding of the link between personal relationships in selecting LSPs and market conditions, political systems, or cultural characteristics. BSRs likely differ in cultures that are distant to the Chinese one, or close ones such as the Japanese or the Korean[1216]. Also, in the explorative expert interviews and the brainstorming phase of the Delphi study, participants shared information that went beyond the scope of this research. For instance, they indicated how Guanxi is built through daily interactions between operational associates, or how customers of a buying company will unofficially appoint specific LSPs. A future study with a focus on these points should, therefore, be undertaken. Moreover, as they were identified as the most relevant factors, panels for the Delphi study were set up according to the types of firms and their degree of internationalization. Comparing results from these diverse panels has helped to gain deep insights into the intricate workings of Guanxi. In further research, other panels, for example, according to firm size, could be set up to gain even more granular insights into the future development of Guanxi's influence on the LSP selection in China. Finally, as Guanxi is a multifaceted phenomenon that keeps transforming and adapting over time, future work could be done in the form of longitudinal studies[1217]. Also in the time to come, China will remain a fruitful ground for locally-based supply chain management research[1218]. As China keeps expanding its global footprint, increasingly participates in global trade, and is moving closer to the world economy's center of gravity, it will become even more imperative to understand local business practices and react to them[1219]. This study shall serve as a foundation for that.

[1215] Kapadia (2018); Buxmann et al. (2004), pp. 295-307; Thun/Hoenig (2011), pp. 242-248.

[1216] Hitt et al. (2002), p. 358; Yang/Wang (2011), p. 492; Usunier/Lee (2005), p. 464; Yeung/Tung (1996), p. 63.

[1217] Fan (2002b), p. 546; Yang (2002), pp. 459-460.

[1218] Liu, Xiaohong (2014), pp. 403-405.

[1219] CIA (2019); Economist (2019); Quah (2011), pp. 3-9; World Bank Group (2018a); Gorgoni et al. (2018), pp. 580-604.